PET Radiopharmaceutical Business

PET Radiopharmaceutical Business

Andrea Pecorale • Maria Carmela Inzerillo
Piero A. Salvadori

PET Radiopharmaceutical Business

An Insider's View

 Springer

Andrea Pecorale
PET Manufacturing
GE Healthcare
Crescentino, Italy

Maria Carmela Inzerillo
Quality Assurance
GE Healthcare
CHIERI, Italy

Piero A. Salvadori
CNR Institute of Clinical Physiology
Pisa, Pisa, Italy

ISBN 978-3-031-51910-9   ISBN 978-3-031-51908-6   (eBook)
https://doi.org/10.1007/978-3-031-51908-6

This Springer imprint is published by the registered company Springer Nature Switzerland AG
The registered company address is: Gewerbestrasse 11, 6330 Cham, Switzerland

Paper in this product is recyclable.

Tell me and I will forget, show me and I may remember, involve me and I will understand.

Kong Fuzi (Confucius)

Preface

Following the publication of the first volume *Essence of the PET Radiopharmaceutical Business: A Practical Guide*, many colleagues and friends asked me whether it was my intention to write a second book in the future; my answer was affirmative.

There were some points in the previous volume whose development did not fully satisfy me.

The first was related to how great operational results can still be achieved, despite the small organizational dimensions of a PET production site.

In this topic, I think that an overview of the differences between the much larger and more organized SPECT radiopharmaceutical manufacturing sites and the PET sites could help the reader to perceive the technical and organizational elements that characterize a PET site and that are the basis of its activities.

The SPECT facility I describe is where I had the opportunity to work in the 1990s and that in some way provided me with a great and useful experience before I started the PET production project, which began in 2001 and still continues to be fully operative nowadays.

It was a professional, but first of all a human experience, without which I would have hardly been able to set up the next challenge with PET tracers and the short-lived fluorine-18 as a radionuclide.

The industrial culture that I have learned over many years of SPECT manufacturing activity, combined with the awareness of the implications of radioactive risk, much higher than in PET radiopharmacy, matured the technical knowledge that allowed me to face the next challenges with sufficient and reasonable confidence.

In the SPECT business, all the technical and radioprotection challenges were, in fact, so impactful and requiring such a detailed analysis, that even the slightest operational changes involved months of studies and feasibility tests.

I was privileged in this; I must admit.

Many if not all the SPECT productions programs I worked on at the beginning of my career no longer exist, at least in Italy, and manufacturing facilities have concentrated in fewer locations; however, a large portion of Nuclear Medicine diagnostics still remains hinged on SPECT products. Younger colleagues, who face the world of radiopharmaceuticals today, mostly focused on PET, may not feel the wonder we received listening to legends, like that talking about the operator who was "able to light a match remotely by using the tele-manipulator system of a five-meter-high, heavily shielded hot-cell."

I would suggest to those who intend to enter the world of PET radiopharmaceuticals to spend first a few months in a SPECT facility; it would be useful to get experience and understand the business philosophy by living in the world where "it all began" and develop the specific mindset needed to face all the operational challenges and their connected radioprotection aspects.

PET market is also very variable, particularly in the context of its organizational aspects; therefore, I felt useful to make a comparison between my PET operational experience in Italy and other European situations, in which I was involved over the time. A picture emerges in which the temptation to operate with the logic of the "silos" (unwillingness to share information or knowledge across different departments), always present at all levels, has fewer chance of success in PET radiopharmaceutical production than other conventional pharmaceutical frameworks.

One aspect, that is particularly significant in my opinion, recalls the "problem of the spider in the hole": how to get it out! That is exactly what critical issue occurs in PET: the need to "pulling out" the production department from its comfort zone and oblige the operators to an open and continuous confrontation with all the other functions, like distribution and commercial departments, not directly involved in the preparation of products.

Of course, I may have expressed a limited vision on this last point, the one based exclusively on the "operational" perspective of a production manager.

The second aspect that, in my opinion, deserved an in-depth analysis is the development of the collaboration with third-party companies that act as contract manufacturing organizations (CMO).

In any business, the innovation and evolution of products remain central. Regarding the development of new tracers, I tried to focus my analysis on the relationship between Research and Development (R&D) and the commercial world as well as the operational mechanisms that should guide the company functions in the choice of new radiotracers.

I thought it would have been useful for the reader to see what, of course in my view, are the main differences between "new," "mature," and "still developing" tracers: elements that I tried to summarize also in a comparative table.

Two "evergreen" aspects, further integrated with respect to the first book, are included in the analysis: order management and production reliability. On these points, I included examples of classifications to promote the logic of comparison and the analysis of results, especially in the context of production networks that include several CMOs.

In the second chapter, I explored the subject of production projects conducted in collaboration with nonindustrial partners, including Academia. This subject is very close to me because it was the frame of the construction of the first PET tracer manufacturing site that I developed directly.

During my professional activity, I have experienced the acquisition from another company of the PET site I managed. Despite the shocking situation, I think that it was a useful experience and an opportunity for me to learn how to explain and report, in this kind of a situation, the peculiar technical/regulatory aspects applicable to PET products. In fact, there is no specific literature regarding PET sites that can help both parties, the buyer and who is acquired on the evaluation of the main aspects to be taken into consideration as well as the connected risks, including the economic ones.

Finally, the PET warehouse chapter: a personal tribute to the small world of PET tracers, often little known, in which night shift work is conducted, and one never knows what may happen.

I think it would be interesting for the reader to have PET views different from what I was able to offer, which was limited fundamentally to the production part in an industrial environment. I therefore involved in the project of this book two colleagues and friends, Maria Carmela and Piero. We have come a long way of working together and shared, each from our side, the adventure of PET tracer manufacturing.

Crescentino, Italy   Andrea Pecorale

Preface

When Andrea contacted me to discuss my possible interest in participating in the drafting of a chapter on PET Quality Assurance (QA), I initially had some concerns.

In fact, for about a couple of years after my retirement, I spent my time doing my best to recover my personal life, and I had only few occasional participations to the activity of a pharmaceutical association.

However, the proposal was interesting to me, and, in the end, I agreed to collaborate, even with a special focus.

I found particularly stimulating both the opportunity, even personal, to carry out a critical and structured analysis with my experience in the QA for PET sites, as well as the hope of providing colleagues, and those who approach the QA world, with some indications, based on real experiences. Hopefully, it could provide the reader with practical methods and consideration which are applicable to routine, everyday activities, taking however for granted the knowledge of applicable pharmaceutical regulation.

It is a sort of summary of what QA has been for me, not just in the PET field, and a reflection on the most frequent mistakes we experimented, that it would be better to know in advance not to repeat them.

To simplify reading, I proposed to use in the chapter the format of an interview, containing a series of specific questions raised on the basis of the curiosity of a representing member of the production team (Andrea). I found interesting that the interview began asking why I chose working in QA and what personality traits are needed in this environment to work effectively and to survive.

Moreover, I added some practical descriptions of the methods I have adopted to recruit and develop the skills of QA personnel in my team, and in the teams of third-party sites as well.

To answer Andrea's questions on some specific aspects of the Quality System (QS), I have provided some general information about procedures, deviations, audits, complaints, and Key Performance Indicators (KPIs). However, I always aimed at a very practical approach, putting to evidence the aspects that I personally experienced, sometimes with unpleasant emotions and sometimes with fun.

Finally, Andrea and I wanted to make together an overview on the aspects that should be focused when checking whether the QS under evaluation during an audit is a real one or it is a movie broadcast only for auditors. It looks like a joke, but if one thinks applying QS requirements only temporarily and not continuously, hoping to save money or simply because the risk was underestimated, the future can turn very expensive and, in particular, not convenient for the company.

I hope the reader appreciates this part as a simple reflection, looking backwards after some time, on a project of construction of a QS, which was developed through gradual attempts, but at the same time had to lead to real effective results, such as production robustness and sale. This reflection led us to conclude that, despite all the difficulties, we had done a good job.

Chieri, Italy   Maria Carmela Inzerillo

Preface

The introduction of new technologies in medicine has always had a fascinating history. What is now mentioned, in the everyday discussion of patients with their doctors and relatives, as a common, commercial device or product has been before a tricky research item in a lab somewhere in the world. In my opinion, it would be too partial a story to limit the description of my experience of managing a site for the preparation of PET radiopharmaceuticals in a public clinical research center, solely to the industrial production of FDG (and other tracers) and neglecting the context in which this occurred.

I felt it interesting to look at my background experience and retrace the underlying history that has encompassed more than three decades. Over this period, the basic PET pillars have widely changed and serendipity played its role.

Indeed, the biomedical landscape is a wide and deep ocean, in which the eddy streams of discovery can bring to the surface new tools and sink others. Acceptance by the medical doctor community has a role in this. The radiolabeled PET tracer [F-18]-fluorodeoxyglucose (FDG) is a nice example of one of these streams. It is also an example of how, after long waiting in the research lab, a very peculiar product, combining complex methodology and highly impacting technology, can find its way to become an industrial resource. In the evolutionary pathway of such a peculiar product, here I try an historical analysis of the variables that influenced its fortune.

Probably PET would not be what it presently is without FDG, nor such a product would be a clinically relevant tool, without the development of hybrid PET/CT scanners, nowadays installed in almost any medium or large hospital. Of course, no industrial effort takes place unless it is an answer to an unmet need. FDG and PET/CT have synergistically driven the business and determined a level of permeation through the clinics that even new products and applications can develop now and in the future. They also have introduced concepts and problems unknown to the "normal drugs" world.

I had the privilege of spanning my carrier along the way in which PET transformed from a research tool to one of the strongest diagnostic devices available to doctors and patients. The management of the PET radiopharmacy, fundamental to the outcome of the PET diagnostic modality, has changed over time and combined enthusiasm and critical issues to ensure economical sustainability and coherence with the mandate of patients' care.

The collaboration and the partnership between the public and industrial world soon entered my experience and still remain a sound perspective to PET business development. However, the dynamics of relationships between the partners and the variables that can play a role are complex. It will be the core of my contribution trying to depict these points by starting from my personal experience, a sound part of which was with Andrea and Maria Carmela.

Although my considerations remain personal and do not yield a recipe, I aimed at focusing aspects that may be encountered and critical situations that may be faced.

In the end, I decided to add some personal stories and commentaries. The intention is to spotlight on specific issues: an episode, a way of thinking, a challenge that may better convey the spirit of the moment. Hopefully, this will add some colour to the core story.

Pisa, Italy   Piero A. Salvadori

Contents

Chapter 1
Management of a PET Site, the Ability to Reach Great Results Through a Small Organization

Contents

1.1 Differences Between the Management of the Production of SPECT and PET Radiopharmaceuticals

The first time, in August 2001, that I saw a PET production facility dedicated to [F-18]- Fluorodeoxyglucose (FDG), I was impressed by the small size of the laboratory, in particular of the production and quality control areas.

I was used to work in a SPECT radiopharmaceutical manufacturing plant, made up of different departments that hosted Class A clean areas, large hot-cells, not to mention the complex technology adopted for the loading of Tc-99m generators starting from absurd bulk quantities of Molibdenum-99 or the automatic systems for the production of I-131 capsules for metabolic radiotherapy. I wondered how it could be possible to carry out an entire manufacturing process in an area of barely 50 m². Quite a different situation from other equipment installed at the research site I was visiting.

The thing that struck me most was the majestic and imposing cyclotron, from its magnet seven bombardment lines departed, ending in three large, shielded chambers. It was not a latest generation cyclotron, it had in fact been manufactured in the

© The Author(s), under exclusive license to Springer Nature
Switzerland AG 2023
A. Pecorale et al., *PET Radiopharmaceutical Business*,
https://doi.org/10.1007/978-3-031-51908-6_1

1980s, the shielding entrusted to rooms with 2-m thick walls. A nice glance which, however, seemed not balanced with the small size of the area intended to produce the radiotracer.

The substantial difference between the two productions, PET and SPECT, consists, in my opinion, in the extremely advanced automation adopted in PET, definitely a necessity, due to the short half-life of the radioisotopes, but also a consequence of the long permanence in the research labs. Instead, in SPECT processes, the radioisotope has a relatively long half-life, compatible with more standard pharmaceutical approach, and an earlier appearance on the market, which somehow slowed the adoption of experimental advanced technologies. It is difficult to change the cards when the game is ongoing.

The radionuclides I-131 and I-125, for example, have a half-life of 8 days and 2 months, respectively. Therefore, isotope manipulation processes can benefit from more time than PET radionuclides, whose half-lives range from minutes to few hours.

In my SPECT experience, the processes were managed by operators who carried out the various steps by means of remotely controlled tongs to manipulate all the devices and components needed in the process, which were confined within shielded hot cells to preserve personnel and environment from radiation exposure and nuclide contamination.

The process steps, including synthesis, purification, formulation, and dispensing, that in PET business are carried out sequentially, and almost completely automatically by synthesis modules, in the SPECT facility relied upon a fundamental contribution of manual skills.

I remember the use of specific heating systems, such as small autoclaves, but nothing that can be close to an automatic PET synthesis and dispensing module.

Obviously, less pushed automation does not necessarily mean that you can "take it easy," in terms of being less concentrated on the progress of the process.

SPECT productions are scheduled daily and on the basis of the orders received, as it is in PET business. And here too, customer' orders get urgently in the department.

But if a production run went wrong, it was possible in general to carry out a second production run, given the availability of a precursor radioisotope stock, which was normally ordered with a certain excess.

The organization of the SPECT process is a sequence of "operational blocks" involving different hot-cells. The direct control of the operator allowed, if necessary, the effective decontamination and restoration of a cell through removal of materials and replacement of starting materials, more quickly than in the management of a process based on an automatic PET module in which the whole process is running in one single environment.

PET synthesis modules often work by using kit-like cassettes. They contain all the necessary synthesis reagents and purification consumables: it is very difficult, if not impossible, to isolate the part of the process that did not work in case of a failure, and it is inevitable that the defective process is aborted in its entirety.

Nowadays, the hot cells hosting the PET modules have no tele-manipulators. The automatic synthesis module would require a too fine ability to intervene.

In the case of Mo-99, the "repetition" of production was not possible, given the high cost of the isotope and the very large amount of radioactivity required to start a production run. However, the chances of failure were very limited, as the process was very simple, in practice the precursor was diluted and fractionated by passing it on solid supports that absorbed it. No more chemical processing, nor there was a synthesis to control. However, stops were possible, they could be caused by mechanical failures of the system that regulates the distribution of the solution inside the resin columns that trapped the Mo-99, such as problems of centering of the dispensing systems. The preliminary tests carried out in the days preceding the production and the possibility of intervening effectively with the tele-manipulators, allowed a high level of manufacturing reliability.

In the production of Mo-99/Tc-99m generators, the radioactivity involved, the size and technology of the plant make the process very similar to a real assembly line.

I remember the arrival of the molybdenum container as an event: in fact, at least two people in addition to the radiation protection technician were in charge to provide transport from the truck (obviously dedicated) to the manufacturing department. Once the first safety protecting devices were released, the container was introduced through an elevator into the cell where the primary container was extracted by means of specific tools and tele-manipulators, which were capable of fully reproducing the movements of the operator's hands.

This phase had something magical for me. The use of tele-manipulators required specific training and probably (given the failures of my few attempts), a natural "talent" that few have.

I remember, it was rumored among the old employees, that an operator was even able to "light a match" using the manipulator in the cell dedicated to radiotherapy sources (five meters high!). Of course, we are speaking about corridor tales, but it can be an indication of the level of skills these technicians should acquire.

During my work in SPECT manufacturing, the experience with Iodine-123 proved to be quite similar, somehow preliminary, to the management of PET radionuclides. This radioisotope, unlike Iodine-131 and Iodine-125, has a much shorter half-life: not days but only 13 h. Furthermore, the radionuclide was not produced on-site but at a cyclotron center abroad, and it was to be used in the synthesis of labeled products as it reached our facility. It was a sort of PET process in which the radioisotope, traveling by road, took several hours to be transferred from the cyclotron to the production department. Therefore, its delivery time in the SPECT factory had to be synchronized with the radiopharmaceutical processing and its distribution tuned to customers' orders. At that time, in practice, the production of iodine-123 tracers was considered the ultimate limit for a commercial radiopharmaceutical. Soon, I would have appreciated that a half-life of 13 h is a huge advantage, as compared to the 109 min decay rate of Fluorine-18!

Shipping Activities

The number of people working in a SPECT department was greater than in a PET facility. The activities carried out by the staff are organized like in any pharmaceutical company; tasks are well defined and roles are more detailed; workers are assigned to a given production step and seldomly take care of other aspects, such as packaging or labeling and shipping.

In a PET facility, it is quite normal that the same production operator is trained on very different activity from the functioning of the cyclotron to the synthesis, the dispensing, and even the shipping of packages. There are reasons for that historically PET has a very small-staffed and limited volumes of dispatched product.

Furthermore, SPECT radiopharmaceuticals are produced in different pharmaceutical forms; therefore, the flow of different products, with different production/packaging methods (radioactive sources, solutions, capsules, generators) requires separate lines specifically equipped with ad hoc machines, each requiring trained operators, ampoules, vial, blisters, either containing liquid of solid products, and, of course, a large poll of personnel in the shipping department. It is a common practice in a pharmaceutical industry.

The radioactive solutions for the production of oral or injectable dosages of I-131, diagnostic I-131 capsules, injections containing Tl-201 or Ga-67, and the low-level activity [I-131]-labeled compounds, just to mention the isotopes of which I had direct experience, were stored, manipulated, and dispensed in separate cells. Products were delivered to customers, upon request, within a time frame of up to 1 week.

A quite different situation was with the capsules loaded with I-131 for therapeutic purposes and the Mo99/Tc99m generators. These products needed to be produced and shipped almost in real time to avoid product self-deterioration, given the high content of radioactivity involved.

The activity of a SPECT shipping department required a constant alignment of production and customer service, giving to the manufacturing process more logistic than radiochemical characteristics.

The constant and intensive handling of radioactive materials in the packaging department—all products converged to this specific site for being shipped—made these operations sometimes even more delicate from the radiation point of view than the production activities themselves.

I remember the product shipment team of the SPECT department, in which I worked, was not a coveted position, being considered instead a kind of "confinement."

At the time, the radiopharmaceutical division was a small part of a very large biomedical group, in which many different production lines and different brands operated. It was therefore possible, even normal, that personnel was occasionally moved to different divisions, all parts of a single company, according to actual needs and market better opportunity.

In this regard, I would like to share with you some anecdotes about colleagues working in the packaging department. Due to the fact that this was considered a merely "handling service" more than a "technical department" by the top management, its staff included the most different people, even transferred from other departments, sometimes because of modest performances. From time to time, quite peculiar situations occurred in this department, that clearly depicted the singular characteristics of both the operators and their supervisor. One of these followed a lively discussion between a rather particular operator and the area supervisor; the issue was on the way of having a loaded forklift passing through a service door. The end of the dispute was a catastrophic result: the damaging of a forklift and the replacement of the warehouse door. This department, just because erroneously considered not enough technical, was poorly kept under internal observation and control. It was mostly run, with "personal" management criteria and methods, by a supervisor who, because of the inhomogeneous environment and its marginal consideration, had developed a rather harsh mood toward both the external functions and the internal personnel, with the result of creating an isolated area, like an island. This situation becomes very evident when the division came under the control of a foreign company. On that occasion, to remark his own authority on the territory, he covered all the windows of his office with the country flag of the new company. Unfortunately, this kind of a joke was not equally appreciated by the HR department, and some reaction quickly got on track. The isolation of the packaging department was soon interrupted and overcome by introducing teamworking not only within the department itself, but with all external functions. Indeed, teamworking and communication certainly were basic in the management strategy, later on, of our PET sites.

It considers that the activity in this area is less interesting than the activities in the production area, and it is difficult to motivate continuously the operators. Working in the radiolabeling of compounds and in the production of generators allowed a certain variability in the activities carried out, while shipment procedures can be very dull.

However, extreme care was needed in matching the customer's request with the right product and dose, and this was tougher in the past than nowadays.

Definitely, the use of barcodes today can greatly reduce the possibility of an error. At that time, I am talking about the mid-90s, the check was completely visual, without any technical support or automation, only by comparing the shipping document with the label affixed on the containers. This operation was not the easiest to perform with highly radioactive products. The label could be read either through a lead glass, as in the case of the solutions dispensed in vials, or through a "quick" look inside the shielded trolley containing the capsules for diagnosis or therapy.

As a matter of fact, it was a work continuously in contact with radioactive materials that required dexterity, a cool head, and excellent eyesight to limit radiation exposure, and prevent any risk contamination and mixing of products.

Radiation Protection

In the contexts in which I have worked, there had always been an internal dedicated radioprotection service. In the case of a SPECT facility, this was an operational necessity.

As I previously mentioned, the high number of people on staff, the use of different isotopes (with relatively long half-lives), and the amount of radioactivity involved, required a capillary control system of the environment and the workstations. Safety procedure extended to the need to wear special clothing, including ad hoc shoes and overshoes, before entering the work area; they had to be used exclusively inside the departments and were dismissed when exiting. Radiation protection and clean room operation mixed together into very severe gowning protocols.

For about 30 employees, two people were exclusively in charge to look after to radiation protection, with a qualified expert frequently present on-site too.

Their job was not easy, in particular because their indications could considerably affect the speed of operations and the daily activity of operators at the various departments. They were often the object of heated discussions, because indirectly their decisions could potentially create production and shipping delays.

In many PET sites, the smaller size of the facility, the use of short half-life isotopes, and the availability of automated and standardized processes lead to a lower radiation risk; as a consequence, the same staff is dedicated both to ordinary radiation protection activities and production. Of course the qualified expert still supervises the overall activity and checks environmental and personnel exposure/contamination track records to ensure safe operations at the site. Therefore, specific training on radiation protection is required for all technicians, in particular on the management of routine operations and emergency events, which are fortunately less severe or even less possible than in SPECT manufacturing. The quick decay of radionuclide often limits the possible damage and even resolves environmental issues.

Working Shifts

The need of organizing the staff on night shifts came only with PET tracer manufacturing.

In SPECT, at least for the production volumes I had to deal with, there was never the need to have personnel on night shifts.

Sometime, there were specific situations: for example, a couple of days a week the production started earlier, at 7:00 a.m., due to the arrival of the "short-lived" radionuclide I123; while, on Wednesday afternoon, the working time was continued until 9:00/10:00 p.m. to complete the production of generators. These situations could be simply managed with a change in staff hours or with a moderate overtime work.

Some difficult situations still could be possible: for instance, when a bank holiday was on a Wednesday or Thursday, i.e., the manufacturing days of the generators, production had to take place anyway. Indeed, it was a different way, in some cases, to celebrate Christmas or the eve of new year, but that's it!

In PET, shift working is a necessity, unless you are producing in-house for your own patients. In fact, due to the short half-lives, it is almost impossible to deliver the product to customers at 7–8 a.m. without starting the process early in the night. Nor it is sufficient to cover the time interval needed with just one shift, when more than one process has to be performed. The working time window in such a case can range from 10:00 p.m. of the day before to 5:00 p.m. of the day in which the delivery is scheduled. Then, the number of people on staff, their allocation of shifts, and shifts organization are challenging.

On-Call Availability

A new experience and a problem I had to face in organizing a PET manufacturing site was the "on-call availability."

The timing of SPECT tracer manufacturing was mostly organized from the morning to the early afternoon, with the exceptions mentioned before. As I said, it was a process with a considerable staff available. PET manufacturing, instead, relies on a small staff. By working at night, all the employees of the manufacturing site, and not only the responsible figures, could be contacted anytime, and well outside the standard working hours.

While the SPECT business could be organized as a pharmaceutical company working with a daily routine, and a planning usually based on months or weeks, the PET production is organized on a schedule built on days, even on hours! The planning of patients at hospitals is very variable, and often slots are completed only on the day before the diagnostic session. This translate into a quite hectic, sometime frantic routine made of an incredible number of doses/destinations to be assigned to batches and batches to be released per day.

Supervisors and managers have harsh life; in addition to often having to spend more hours than expected in the normal working day, due to the characteristics of the business and the small number of people, they must be available for night contacts in case of problems.

The training of personnel is, of course, of paramount importance in reducing these events to exceptional, and a big effort has always been done on this; however, the "human" aspect and the psychological discomfort that these situations entail for the employees cannot be eliminated.

Thus, the quality control technician, facing an anomalous result or an instrumental problem in the middle of the night, could feel the need for a direct and urgent discussion with the supervisor. Likewise, the supervisor or the QC manager will need to know that something may go wrong with the production and the lot be refused. It is obvious that, upfront the possibility of interrupting production and having the patients untreated, some sacrifice is accepted.

Similar situation can be faced by production technicians on duty. In some cases, it is possible that they will be contacted to rearrange the production schedule in the middle of the night, or even requested to return to the site to rearrange shipments or start an additional run: patients come first!

For the same reason, employees can be requested to accept modifications of the already scheduled shift, even by many hours, or be available to work on holidays or weekends.

Such a compelling situation raised the need to define an "on-call availability" system. This was not only dealing with work organization, but necessarily had to be managed in terms of an economical compensation; thus involving human resource (HR) skills and competences.

First of all, an initiative was set up to increase the remuneration of worked hours in a nighttime slot, typically from 10 p.m. to 6 a.m. In this way, the first problem was effectively addressed, by prioritizing as much as possible the manufacturing night activities.

In parallel, a compensation was set up for an on-call weekly availability.

The number of on-call weeks assigned to each employee was fixed a month in advance by the site manager. The variables considered were as follows:

- The number of interventions requested to the employee in a specific interval of time
- The calendar (number) of shifts scheduled for the month
- The expected presence on duty of the employees, thus considering vacations and other planned activities.

The relationship with HR to define these principles and time organization for the on-call availability was not easy, since the PET activity was new and developing, and forecasts were hardly predictable.

The starting point I used to define the on-call availability was the frequency of events that required intervention outside the standard hours.

It was not easy to have a plan, because the origin of the "need for intervention" was mostly linked to technical aspects, such as failures of equipment, that could not be foreseen in their occurrence without an historical analysis, and we were just at the beginning.

The fact that the equipment is subjected to a regular maintenance process, in accordance with the indications provided by the manufacturer, does not constitute a 100% guarantee of the manufacturing continuity of a PET site.

In addition, the PET process takes place over a short time interval, the crude manufacturing process ends in a couple of hours; nevertheless, it involves many steps and different instruments and apparatuses; each of them can trigger the anomalous event or be the origin of a failure. Any of these can occur and statistically increase the production risks.

When I started my PET experience, I had not considered (and expected) the necessity of an on-call availability system. Likewise, the company had not considered an overcompensation of the employees that would have been operating, beyond the usual routine, in a new environment that included new activities and a new stressing organization of the work.

When, many years ago, I first initiated the PET manufacturing activities, unlike the current situation in which there is a back-up network participated even by all competitors, I could not count on any external support in the case of a production failure. The patients would have missed their treatment, and this would have been very sad for them, for all of us, and also critical for the relationships with customers.

Therefore, the lack of reference data from historical manufacturing records, and the inability to have a back-up of failed productions, made the problem of on-call availability extremely pressing.

I remember that, like me, no one in the company was prepared for such a situation, least of all the HR department.

The national labor contract had already included provisions on night shift salary increases, but the employees were not satisfied with that: the failures were occurring quite frequently, despite the efforts made by the company that heavily invested in maintenance activities and corrective actions.

I got caught between two fires: (1) those who, despite being passionate about their work, were overwhelmed by a non-stop process, and (2) those, used to the SPECT manufacturing and its minimal rate of abnormal events, who were not comfortable with PET manufacturing working organization and its frequent anomalies.

In this frame, I decided to approach directly the HR department, involving also my collaborators in the discussion, with the aim of reaching a shared and clear picture of the current situation.

I understood that it was necessary to make HR aware of what happened daily in our laboratories, and explain them the very different situation with the new business (PET), as compared to the old, consolidated historical view (SPECT) existing in their offices.

At least 4 days of intensive visits were needed, in which my collaborators and I spent a lot of energy in describing and showing our operational steps, the difficulties and the criticalities that may be faced to HR colleagues. In the end, they even had a direct experience with the participation in a night production run.

In this way, we could reach a uniform base of knowledge of the business and share a common language, in which there was a clear reference to equipment, processes, working shifts, abnormal events, and overtime work needed to face unpredictable problems. Finally, this leads to the definition of a fair economic compensation, and a full satisfaction of everyone.

There was also an emotional involvement that the negotiation process was able to generate and, as a proof of the positive results reached, 7 years after the closure of the early pilot plant, I met old HR and PET colleagues, who still remembered the phases of the negotiation, and recognized themselves as architects and protagonists of a moment of absolute novelty in the life of the company.

The gross problem was solved, however the risk remained somehow on hold, because overcompensation cannot resolve the shortage of personnel or an undersized team in cases of excessive workloads. In real life, it was not possible to negotiate the commitment or availability of the employees beyond a certain limit, even if they were very skilled and highly motivated. In such a case, it was necessary to consider the possibility of increasing the resources.

1.2   An Italian Technician in Europe and in the World: What of our Way of Managing the Production of Radiopharmaceuticals Can Be an Excellence? When the Imagination Can Be an Advantage

During my professional activity, I often had the opportunity to visit manufacturing plants located in other European countries.

In fact, I was lucky enough to be the first within my company to have developed, in Italy, a PET pharmaceutical facility.

Therefore, it was quite normal that I was called to participate in similar projects that were developing within my company in other European countries, like Portugal, the Netherlands, Norway, and the United Kingdom, and provide operational indications and suggestions to the new upcoming PET manufacturing sites. On these occasions, I always asked to attend night productions for a week or, in case the plant was not yet in operation, to simulate "dummy" night productions for several days.

Direct experience is always better than only listening: it was not a surprise I detected sometimes recurring problems, that I try to summarize below, that recalled what I already encountered during my early experience of starting the Italian production sites.

Personnel Organization

- Staff organization is numerically suitable to manage the project pilot phase, but is insufficient to face a routine production with the necessary shifts.
- Personnel recruitment is defined without taking into account the need of night shifts in the full operational phase.
- Production times are not sufficiently optimized, due to poor alignment between the various departments, especially production and quality control. For example, not considering the need to start QC activities as soon as possible (e.g., by using the first dispensed vials, without waiting for a completed dispensing). The preparation of the material and documentation necessary for the various activities was not arranged in advance, wherever possible, so as to facilitate the performance of the night staff.
- In one case, the shipping activity was not considered as an integral part of the process, so that the preparation of the shielded containers and the transport

documentation were carried out only after the completion of the production, with an unavoidable (and unacceptable) delay of the shipment and delivery time.

- Not sufficient time and space were dedicated to the management of returned shielded containers and their sanitization.
- Procedures and timing of shipping activities were not efficiently and clearly transferred to the PET production staff. This occurred, in particular, at sites dedicated to the production of many tracers, they were more complex and the shipping department was not specifically dedicated to PET products.
- Poor alignment existed between the various departments, like production and distribution, especially in case of production problems and delays.
- A clearly defined and shared work plan, in terms of staff activities and times aimed to respect the scheduled shipping time, was missing.

Maintenance Activities

- There was a poor technical knowledge of the equipment and a consequent inability to carry out minimal maintenance, without requiring the intervention of the supplier.
- A stock of the spare parts, which were more frequently requested for urgent replacements, was missing.

Interfaces Among Departments

- A lack of knowledge of the margins of tolerance accepted by customers, in the event of problems or insufficient production capacity.
- Procedures and emergency contacts, in case of need for night product back-ups (where applicable), were missing.
- Insufficient awareness and daily verification of the real production capacity, in the drafting of the production plan.
- Production times that were not perfectly aligned with the delivery needs of customers: e.g., in-house decay of the activity due to unnecessary storage of samples before being shipped.
- A clear plan for the direct communication with customers was missing; especially in the case of production failures and/or the need to reschedule the delivery of doses. In this specific point, especially in the case in which the new PET site is part of an already experienced production structure, such as the production of conventional drugs or SPECT radiopharmaceuticals, the difficulty arises from the need to modify processes and operational habits that have been consolidated over the years.

- Poor/missing communication between the commercial department and the production site staff about the actual requests of customers and the commitments taken with them.
- Clear contact references with the transport companies, particularly in case of emergencies, were missing.

In-Bond Shipment

- Applicable operating procedures were sometimes not functional for a timely delivery of product to customers. In this regard, I suggested the use of the "in bond" shipping procedure. In practice, the product could be shipped before the completion of all tests, but any possible use of the product was inhibited, even if arrived at destination, until it was released by the QP and clearance was given to the customers for use. This saved time, because the product was already in transit while QC tests were ongoing; the analytical results were transmitted to the customers, usually at arrival or before, in time for the planned treatment of patients.

Training

- Often procedures were not fully understood by the staff, thus leading to a slow-down in the overall production chain.
- The personnel was not sufficiently trained in managing the dispensing system. This is in fact a very critical equipment but often the only resource available to adjust or modify a scheduled production plan that needs a change, such as the distribution of activities to customers, and variation of the dispensed volumes.
- Personnel was not fully trained on how to prepare and manage the shipping containers (shielding containers and plastic Class A shipping bags).

In general, all the new sites were developed with a "silos" mentality, where the ultimate goal was to carry out the required production activities, with poor vision on customer's need.

The aspects that surprised me the most were as follows:

- The lack of on-site contact references for the prompt communication with customer in case of a production emergency
- The evidence that the production failures were not perceived by the staff as an event to be immediately faced and treated, with all the necessary resources.

I remember the case of a failure, due to an out-of-specification (OOS), in which the technician informed me in absolute calm and relaxation that the QP was rejecting the batch. The thing that struck me was the apparent absence of any emotional involvement in transferring, to an external observer and monitor, this information.

It was clear to me that the event was felt by the employee, who probably had completed his job, as something "that may happen" and not as a serious lack of customer service.

The fact that the batch was not released and that some patients would not have received the dose, seemed an issue that did not affect the operator's conscience nor his professional responsibility.

I remember that the QP had just left when I asked to be able to discuss the results with all the involved staff and verify the possibility of rescheduling a new batch. Unfortunately, there was no operational plan that considered such a possibility, not even the prompt communication with transporters and customers.

In the event of a failure, like the one I saw, the procedure they had was to send a communication to the customer service, while the analysis of the failure was delayed to a later time, and the possibility to reorganize a new production batch even was not considered. Indeed, it was a situation of poor consideration of patients' needs, but also a limited vision of business future fortune.

The Italian PET sites, with longer experience than that site in which the problem occurred, had already matured prompt reactions to these possible critical situations, and had put all available imagination in solving the problems as they came: years of life within the nuclear medicine world had conveyed the message that our job was not only satisfying customer requests, but primarily ensure patients' treatment.

When I mention imagination, it is not a futile thinking. I remember the case of a dispenser, used to divide the bulk tracer solution into vials, which had problems: the jaws of the mechanical arm, installed by the manufacturer to move the vials across the various working stations of the automatic dispensing unit, sometimes did not correctly gripped the aluminum cap of the vials. The result was the falling of the vial on the floor of the hot cell, in which the dispensing unit was installed, and the impossibility of retrieving it. It was a rare event, probably due to minimal differences in the tolerances of the aluminum caps: but it was too much for the jaws.

The simplest solution, in this case, is to contact the supplier of the equipment. This usually requires an ad hoc study by the manufacturer's engineering service and modification of routine spare parts. Usually it is an issue that yields reluctant reactions from the supplier, expensive prices after repeated pressures, and a very time-consuming track.

Therefore, before moving in this direction, since the aluminum jaw was a component not intrinsically connected with other components of the system, as all the teams were willing to speed-up finding the solution, we decided to turn our mechanical problem to a local, very well skilled, mechanical craftsman.

The craftsman listened to the description of our process with great interest and started immediately to create a CAD model of our component, from which he obtained different versions of the modified jaws, which were immediately tested in our dispenser, of course outside the usual production activities. The result was amazing and led to the definitive elimination of the problem in a very short time.

What I wish to remark with this anecdote is that one failure of even one batch was perceived by all the staff as a serious event, that had to be promptly solved even using creative ways.

This approach was the consequence of a training activity, which was aimed at the awareness that the treatment of the patient depended on everyone's work. Everyone felt responsible of any failure and was committed, working as a team, to solve it.

However, imagination and emotional involvement may not always be a positive spring, and the rationale Anglo-Saxon approach was a teaching example for me.

The right calm in the decision-making process should never be lost. It gives you the time and the opportunity of a rationale thinking and can be a useful support for deciding whether an idea, presented at heart as obviously true, is the right way to go or it is a too less pondered solution, and cause instead delay in finding the real solution. Indeed teamwork and expert advice can help on this.

1.3   Commercial PET, "The Spider Out of the Hole"

Until the moment I started dealing with PET business, the commercial team remained always a bit far from my routine activity in the production department.

Of course, there was no lack of direct comparisons, but this was mostly limited to occasions in which important changes were discussed, or customer's reactions/complains analyzed. For example, it happened when it was discussed the design of a new dispenser, dedicated to the production of I-131 therapy capsules, which could finally accommodate a change of software to be able to accept last-minute modifications of the capsule production schedule. The new dispenser version allowed changes beyond what was before considered a rigid order deadlines.

Indeed, solving the lack of flexibility of the dispenser software closed out a long-lasting controversy between commercial team, exposed to customers' pressure, and production, that had to reorganize the process consequently. Other moments of discussion with commercial colleagues were on the management of complaints, especially when they were allegedly coming from the same problem or issue.

In the more conventional SPECT manufacturing organization, commercial people were quite far from the production dynamics and mindset. For example, the time needed to conduct the analysis and investigation on the root cause of failure and be able to remove it, were less important than putting the effort into having the most immediate customer satisfaction.

This last point should have received priority attention, even with a prompt product replacement if needed, and "without stressing the customer too much" with a calendar disturbance due to the delay linked to (always too long) investigations of the event occurred.

However, in SPECT tracer manufacturing, the processes were simple and production problems were quite rare; therefore, one could work for weeks without any need of a direct interaction with customers, and thus their response—and their commercial interfaces as well—was more pressing in case of a failed or delayed delivery of the product(s).

I was nicknamed the "spider in the den." It was coined for me, due to the specific SPECT contest situation in which I was working, fully embedded in the production area, with almost inexistent external communication outside of production environment.

Starting the PET tracer production, due to the peculiarity of the end-user organization, the manufacturing process, and the radionuclide extremely short shelf-life, the situation radically changed.

A natural consequence of the start of PET production was a big change in the communication process, especially in the communication chain among production-commercial-customer actors.

I realized the need of this change very early. Since the moment in which the management processes of the order placement were being established; in fact, in defining the communication flow of production plan preparation, the role of the commercial contacts was immediately considered, with the specific activities listed below.

1. Collection of product needs in specific commercial (geographical) areas and verification of the effective production capacity of the site to satisfy them.
2. Immediate communication with the customers following any lack of production capacity or delay on the expected supply days.
3. Mitigation of the impact on customers in case of production failures or delays, e.g., with a timely organized, replacement delivery of the product.
4. Establishment of a direct, continuous contact that could reassure the nuclear medicine department about the staff commitment in solving technical and logistical problems.

The prompt contact with the customer is also essential to address any possible legal consequence related to tender/commitment obligations. Moreover, we should consider that back-up suppliers, that in principle could mitigate the impact of failures, may not be present or be able to timely reach the specific customer location.

Following all these considerations, the concern of establishing very close contacts with the commercial people on the field was clear in my mind. In particular because these colleagues, who were generally trained on performance and characteristics of nuclear medicine products, were not yet aware of the technical criticalities that could impact the performance of a PET manufacturing site.

It is clear that I could not expect to find easy justification or acceptance for failures. My previous experience taught me that when company image damage and financial interests are at risk, the level of tolerance gets significantly low.

Then, the goal was to find, at least, a uniformity of language with the colleagues of the commercial service on the field, in such a way that the first communication provided to the customer was sufficiently detailed so that both the reason of the production failure and the indication on its expected restoration could be readily shared with the end user.

It was imperative to involve commercial staff colleagues, as well as customer service function, and train them on understanding what the critical points were.

With this scope, in our organization, comprising an articulated nationwide production network with multiple sites, I organized a three-day national internal meeting of involved commercial staff, directly on the premises of the manufacturing facility. This allowed key personnel, i.e., both commercial and technical staff, to have access to the same information on PET production: the problems that can occur, the available solutions, and how the information can be shared with customers, in order to build reciprocal collaboration and mutual trust.

Indeed, the most frequent communications at the site take place in the early hours of the morning, when the production process is ended, and usually involve three parties in addition to the manufacturing staff: the customers, the commercial contacts of the involved location(s), the customer service function.

There is a vast anecdotal bibliography about contacts with customers. It is obviously depending on the degree of mutual understanding and personal ability to refrain frustration and, sometimes, anger at destination sites, where radiographers and doctors have to deal with severely ill patients.

Never, in other experience of mine, have I been so openly and sometimes violently criticized.

The temptation to react instinctively, however, never crossed my mind. This is because to me, as well as to my team, the target of our work beyond the walls of our facility was the patient.

We figured out the disappointment and the difficulty of both nuclear technicians and doctors, who were obliged, in case of a failing delivery, to manage the situation and reschedule their internal plan for the scanner and the patients, and, obviously, the discomfort for the patients, often very ill, perhaps coming from afar and accompanied by family members, forced to delay their treatment.

Once, the quantity supplied to the hospital was lower than that requested; the process runs with lower than usual yield, and this lasted for several days, despite our efforts to solve the situation. Finally, the doctor was so disappointed that started shouting at anyone in the production staff and the commercial people before any explanation or apology could take place. I never had such an experience while working with SPECT tracers.

The punctual information about any problem to commercial service in the various territories is another important aspect. It is essential that they are informed immediately as they are the primary reference for the customer. In fact, it is almost certain that they will be promptly contacted by the customer in case of supply problems (either a delay or a cancelation).

There is also a general consideration on the effect of not keeping commercial service updated and informed on supply problems: it can give rise to negative beliefs and doubts in the customer's mind about the organization and the professional capacity of the company. Finally, commercial relationships are often bound by tender conditions, and the contract can also be impacted by repeated failures, or these lead to possible contractual penalties and legal consequences.

1.4   A Good Practice on Communication to Customer

My previous experience in SPECT tracer manufacturing was anyway useful in managing specific technical problems that may arise when working with such an atypical drug like those containing a radioactive stuff, and can be found, even harsher, also with PET products. For example, when the activity content per vial results to be lower than that declared on the label.

In the case of SPECT, the impact on the total activity of any process delay is reduced, due to the longer half-life of the radioisotope and the simpler filling procedure. Obviously, depending on the skewed amount, there may be a process deviation and as such has to be managed.

The substantial difference with PET radionuclides is that the problem with vial true activity can be much more frequent: a delay of just 20 min, during the filling process, for instance, can greatly impact the amount of radioactivity actually supplied.

As a matter of fact, the manufacturing facility must have a metering device (activimeter, ionization chamber) that is certified by the national metrological authority. While certified standard sources are available for Nuclear Medicine SPECT radionuclide, the same becomes complicated with PET radionuclides, such as the 2 h half-life fluorine-18. In practice, a complex and expensive procedure needs to be followed to ensure full certification of the metering device. Anyhow, the reading of the true activity leaving the site is always correctly measured.

The values of the doses (activity) contained in each vial and supplied to the customers are recorded and printed on a label that provides immediate evidence to customers.

Unfortunately, it is not always possible to guarantee the same certification process of the radioactivity meter at customers' site (hospitals); therefore, some discrepancy may be found between the activity stated on the label and that measured by the customer. In case of dispute, it could be difficult to compare the two data and share technical clarifications.

To make things even worst, there are multiple labels applied to a packaged radioactive product: in particular one is on the primary container (the vial), and one is on the shielding container (3–5 cm of lead or tungsten) in which the vial is placed to make its external exposure compatible with shipment and handling.

A label reporting the product information, in particular the activity ordered by the customer, is printed during the early phase of the process and applied to the vial before dispensing. In case, there is a delay or a problem during the production (the tracer undergoes a full process comprising synthesis, purification, and formulation), it is possible that sometimes the activity actually dispensed could be different from the nominal value indicated on the label of the vial (primary container).

For any radioactive drug, the activity content must be ±10% of the nominal value as stated on the label. In a normally performing process, the difference falls within this acceptance range. However, it is possible that the dispensing plan is modified and the radioactivity actually dispensed per vial is different from the nominal value indicated on the primary container label, i.e., outside the 10% acceptance range. For example, if the process yield is lower than expected, the activity per vial can be reduced in order to supply as many customers as possible, at least with a dose that allows the doctors to start treating patients and to receive, at a later time without changing their schedule, a compensative amount obtained, for example, via a back-up or a repeated production run.

With the longer half-lived radionuclides used in SPECT, there was much more time to fix the situation and adjust the labels (which are an integral part of the product and respond to regulatory obligations). Due to the time constraints of PET production, the issue (that may have ethical, commercial, and regulatory implications) required ad hoc solutions. In practice, it was essential to consider the following points:

(a) The vial is labeled before dispensing, with the activity ordered by the customer; the external shielded container is labeled after dispensing, with the actual activity dispensed; as a matter of fact the latter is the label only normally read by the customers' operators due to radiation protection regulations.
(b) If the actual value dispensed is within the 10% tolerance of the vial-label, i.e., coherent with the order, the label with the actual value is printed and applied and no action is taken.
(c) If the actual activity content in the vial is higher or lower than the 10% tolerance, even if the shielded container label reports the real value, it was good practice at our site to provide the customer with further information, by a separate written communication or with an explanatory note added to the accompanying documents normally signed and retained by customers.

Whichever way you choose to highlight the possible discrepancy between what is reported on the labels on the vial and the shielded container, it is obviously very important that it can be detailed within the site quality system and based on an approved method.

The health authority approved our specific internal method during an inspection at our facility.

The issue of a discrepancy between activity values printed on the labels and its communication with the customer emerged unexpectedly during an official inspection, and made me near collapse fright, since we could not guess whether the inspector would have approved our method.

This frightening situation came up because the inspector expressly requested to watch the entire production process. On that specific day, the batch shipping was planned at 6:30 a.m.

Thus, the inspector's access was scheduled at 4:00 a.m., to give him the possibility to follow the final production phase of the F18 radioisotope at the cyclotron and the whole manufacturing process of FDG (2-[^{18}F]fluoro-2-deoxyglucose).

To avoid interference with the operators, the inspector agreed to watch the entire process from a glass window.

Upon completion of the synthesis, the operator noticed that the radioactivity in the collection vial of the bulk product was 30% lower than expected. The feared "low-yield process" occurred just during the inspection.

The vials had already been labeled and placed inside the dispensing hot cell which was already closed, with inside the finished product, i.e., quite a lot of radioactivity.

We therefore had to explain to the inspector how we would manage the "deviation" and the logic adopted: (1) a new distribution program was applied, which was based on traveling distances and customers' priority, and (2) the labels of the shielded containers were replaced by new ones, reporting the actual activity dispensed in the vials.

Luckily, we had foreseen and implemented a procedure to manage this kind of a problem, which consisted also of a form placed in the shipping box of the shielded container, in which the qualified person notified and certified the new activity dispensed in the vials to customers.

The inspector was happy with the way the team had approached the event and the documentation system we used. In the meanwhile, I had lost 1 year of life.

Chapter 2
PET Third-Party Collaborations (AP)

Contents

2.1 Development of a New Tracer: The Connection Between Research & Development (R&D) and Industrial Production

In general, when a new PET tracer comes to the horizon and its process is to be developed, the key players, in addition to R&D, are represented by all the functions that are normally involved in the site management, namely quality assurance, regulatory affairs, technology transfer, finance, sales representatives. However, especially if the new tracer has different clinical targets from those previously developed, it is possible that new commercial representatives enter the PET field for the first time. These new persons, in general, lack the basic experience matured by their colleagues already involved in the existing PET productions; therefore, it is always necessary to address them to a specific training on short-lived radiopharmaceutical production and on issues around process completion and product distribution.

They assume a leading role in defining the economic framework and, consequently, they will be the real makers of the profitability strategy of the production sites through their work on the market.

Usually, the commercial agents operating in the field of nuclear medicine have focused their experience mainly on SPECT drugs. Their longer half-life allows for a greater flexibility, in logistical terms, when compared to PET; for instance, they allow the possibility of centralizing production only in a few centers, as well as to organize an effective distribution even over long distances, and in general better costs standardization and planning.

In their role, they will face a number of new difficulties consisting in the fact that they have to support the production and the distribution of the candidate radiotracer that, at least initially, could be less profitable than consolidate tracers. In this task, they are involved, though indirectly, in the difficult company target to attract the existing network of manufacturing sites, that may include many Contract Manufacturing Organizations (CMOs), to enter the new tracer project. In particular, CMOs should decide whether produce a higher volume of doses to balance the low selling price of the new tracers which compares negatively with the consolidated high value tracers produced in the established routine.

This last point, along with reimbursement and sale, should be considered into the commercial strategy and may be not comparable with the previous state of play. The approach that was previously adopted with consolidated tracers, based on CMOs high remuneration for a somehow limited number of doses produced, could conflict with the new situation, that would involve the same remuneration but with the need to allocate more doses and resources.

Often, even earlier with CMOs, the discussion is inside the company and could even generate discussions on the strategies previously defined with the CMOs, in particular their availability to be involved in the production of new molecule and the new business opportunities that follows.

One way to solve this conflictual situation is to establish, at least with the main CMOs, an open line of communication, not considering only the short-term requests but including also an open approach to new lines of research, for both clinical and marketing studies.

This approach would allow to negotiate new operational and commercial needs with a different perspective of mutual growth. However, we had experience of some obstacles that can be encountered along the way of its application, in particular, those due to the reasons briefly listed below.

- R&D has a complex structure; it includes several teams that often work in parallel on different products and are focused on pilot production only. For obvious and valid reasons, R&D is concentrated on immediate chemical, technical, and clinical aspects, rather than on the phase of technology transfer of industrial production and the effective distribution of tracers to various clinical centers.
- No robust lines of communication are usually in place between R&D and the PET production network, including CMOs. R&D is involved in some discussion with CMOs only when problems arise on doses distribution aimed at completing the clinical trial phases.

In general, the hospital centers having a radiopharmacy in their organization are preferred to conduct the early phases of clinical studies, and cooperation and interactions can be quite efficiently managed. The problems raise when advanced clinical phases are reached, in which product outsourcing become necessary due to the need of a larger production volume and distribution effort.

My opinion on this point is to try, as far as possible, to develop an operational knowledge also within the R&D team, by allocating specific resources to develop a PET planning culture. The methods of managing the doses, especially in phase 2 and 3, have many similarities with the management of commercial doses, with the addition of process unpredictability and the need to define a logistics that is often customized and adjusted on a case-by-case basis.

The possibility of involving the commercial team in this planning could be feasible, but, in my opinion, it would be preferable to have a dedicated planning staff with a close connection with both the scientific (R&D) and the involved commercial teams.

- It can be difficult to plan the actual marketing phase with accuracy, as it will depend upon the regulatory aspects, and on the timing of recruitment and onboarding of the various clinical centers involved. A critical task that may delay the completion of the studies, that usually happen to be longer than expected.
- In the development of the clinical trial phase, the recruited centers involved can be in countries other than those in which the commercial PET business has been already developed or is expected to be improved soon. It should be considered that this situation makes difficult to create synergies with the major stakeholders in the world of CMOs.

In the ideal condition, the clinical study should be developed as much as possible within the commercial PET production context that has already consolidated in the various territories. The more this alignment exists, the more it will be possible to have a benefit from it in the subsequent marketing phase. This approach could have the scope to use the knowledge of these sites both in terms of technical skills and experiences of production and distribution of tracers; both are useful for clinical trials and routine activities.

- Nowadays, it is very frequent for a CMO that R&D projects for clinical studies benefit from relevant financial supports, that lead the company to face an increase of investments. In fact, since the CMOs are aware to have the scientific and organizational ability to develop a production that is not yet standardized, they obviously tend to increase the costs for both the technology transfer and the production batches.

In the end, this will also have an impact on the future commercial phase: it will be in general extremely difficult to agree on a reduction of the batch cost and rebalance this price with the actual volumes and sales values.

In the event that a CMO is used as a site for the development (R&D) of the new tracer, and afterwards for its production and commercial distribution, it would be wise to define in the contract both the maximum price and the operational availability (number of batches, days/shipping time) applicable to the commercial batches or, at least, to agree upon the possibility of future adjustments of the costs, following the technological developments of the processes after the experimentation phases.

It is easy to imagine but quite complex to apply, in fact, detailed technical and economic information are often not available at the time the initial R&D project is started.

2.2 The Selection of a New PET Radiotracer

In my experience, the choice of a new tracer is primary linked to novelty and benefits for patients in the clinical practice. I believe that the accusations that multinationals are focused only on mere economic interests are not true. The cure of patients remains central, and I had many evidences on this in my everyday life based on a close relationship with the cure of patients and their needs. However, economic considerations do have a play even during the development phase of a radiotracer.

The strategy to allocate the projects of new tracers within a specific production network or site attains to a wide span of motivations: should we prefer CMOs or a proprietary site? In my opinion (and experience), the following points of analysis must be considered.

1. Competition with tracers or other diagnostic methods already on the market that can reduce acceptance.
2. Marketing strategy, in terms of sale price and reimbursement of the tracer in the estimated period and area of marketing.
3. Ability of the CMOs (site) to face the planned productions, in terms of manufacturing frequency and functional shipping time.
4. Definition of the batch price, taking in account the competition with new tracers that are going to be developed and commercialized in the meantime on the market.
5. Development of the manufacturing process in a way that facilitates the synthesis of the final product, in terms of reagent needs and overall process time.
6. Development of a process, that could result in a high synthesis yield, and will grant a sufficient quantity of tracer to address customers' need, possibly reducing in the meantime the number of batches with less impact on production schedules.
7. The possibility should be considered to use multi-dose vials instead of single-dose vials, thus reducing the product dispensing time, the number of shielded containers, and the handling of packages through the production facility till the shipping phases.
8. Preference should be given to diagnostic solutions that guarantee effect/outcome by a single dose, because of the following reasons:

   (a) Repeated treatments have a more difficult development phase and regulatory process, sometime too long to be faced

(b) A second dose on the same patient obliges the manufacturer to dispense more than doubled activity
(c) The need to inject multiple doses in the same patient causes complications in the logistic of shipment, both for production time and patient's treatment plan.

Operational Processes in Defining the Business Case of a New PET Radiotracer: How to Make a Decision?

During my work at a multinational company, I was able to realize how much the preparation phase of the business case can be complicated and rich of variables, especially in the advanced phase of a CMOs network and the increasing development of new tracers.

At the beginning of my experience of building a PET network, I started from one tracer alone. The approach was very simple: I had to identify the territories of the European countries characterized by a high population density and an adequate number of PET scanners; once I had this map, I selected the production site that was able to serve each territory.

This method was indeed very plain, since the selection was made even before having done a detailed analysis of the market and having received the approval of the marketing authorization (MAA), which is country dependent; however, the method showed both positive and negative elements.

Positive Element

1. Cooperation and network solidity

   The CMOs network is created long before the product marketing phase.

   Each CMO has the opportunity to develop deep technical skills, even before the approval of the product MAA.

   The local team, mainly commercial, encouraged by the presence of a production site that already exists in the territory (even if not licensed and therefore not operational), manages to promptly develop its strategy, by analyzing the commercial opportunities directly on the field and considering the detailed logistical aspects that will ensure the delivery of the product to the final user (in a real situation and not a theoretical project).

   In the event that the MAA has still to be obtained, the local team becomes an integral part of a common project, which can be strongly supported by the corporate, to generate a unique vision of the project, which, as a matter of fact, will provide the opportunity of a shared growth in regulatory, quality, pharmacoeconomic, and marketing areas.

Negative Elements

1. MAA approval process, or necessity of further modifications

    When the MAA comes at a later time than the start of the network, the final MAA contents may require a modification of the process developed by the CMOs, with the possible need to introduce technical changes and/or upgrades (example to allow for a specific dilution/formulation of the finished product). Consequently, it can have a negative effect on the timeline planned at the beginning of the project.

    The impact on the commercial expectations and CMO(s) is also unavoidable. The last could have difficulties in introducing the modifications or even might lose motivation. The presence of alternative molecules, which may result in a competition with the product under implementation at the site, could further contribute to this demotivation, because it might get back discussions otherwise already closed. To keep the interest of the CMOs alive, it could be wise planning technical and test batches. This activity, in addition to improve product knowledge and gain experience on production plans from ad hoc simulations, would be a stress-test of the production capacity and get information on the ability to serve as many customers as possible.

However, time does matter. Therefore, the lengthening of the return of investment timeline, together with possible changes in market conditions, may even lead to a complete transformation of the initial project.

The possible consequences of such a situation could be either the shrinking of the site network or an "on-hold" position for further investments in some new territories. This might last until the achievement of a better development of the market, that could be considered sufficient to guarantee a profitable business.

My personal consideration is that, as far as possible, the presence of at least one production site per territory, country, or macro-region is essential to start a PET project. There is no possibility of developing the knowledge on the radiotracer and attracting customers, without having the real capacity of manufacturing and marketing the product in the same area. This added value, in case of a PET product, is really perceived by the customer, who perfectly knows the advantage of "zero kilometer" connotation for this category of products. Furthermore, the customers will seriously consider our commercial proposal only if they recognize the name of the production sites, possibly having already produced other tracers for them.

It takes a lot of foresight, even financial, to support these projects. If one is not used to PET production, he/her may be tempted by the idea of postponing the project until conditions are more promising, instead of starting to prepare the market even with a limited distribution.

I can confirm that the PET market can only progress if the customer has the perception of the "proximity" of the product delivery. The development of production, its distribution, and the real reaching of all customers can take many months or even years.

When regulatory, technical, or economic constraints delayed the development of a commercial network aimed at PET production in a specific territory, the solution that was usually proposed was to begin seeding the area. This was done by spreading the knowledge about the product across the centers in the target territory; for instance, by promoting clinical studies with the candidate tracer, even if produced in other countries.

This approach was reasonably simple for SPECT tracers, while the PET tracer, with a shelf life of maximum 12 h, makes it very challenging. Indeed, we faced many hurdles, which are summarized here below:

1. A major point to resolve is the competition with the needs of the country where the tracer is already produced and distributed. Distribution in a different country requires the use of a high amount of radioactivity of the batch produced, given the greater distance of the final user, therefore the amount available to the country of production may be reduced.
2. The operational processes (e.g., order management, financial flows) needed important adjustments, such as the creation of interfaces between different countries, which are not always easy to establish.
3. Regulatory, quality, and fiscal aspects had to be analyzed and compelling requirements applied.
4. A long distance between production and user's site normally involves the delivery of the product at times that are poorly compatible with the patient routine that is normally scheduled at a nuclear medicine, usually starting early in the morning and closing in the early afternoon.
5. Very high logistical costs should be considered, especially if charter flights are to be organized.
6. The longer the distance, the greater is the risk associated with product distribution: a shipment that exceeds 3/4 h has a greater risk of not being delivered on time; there are multiple variables, traffic on the road, and potential accidents that can be encountered along the way, transfer between carriers (e.g., transit at airports), boundary crossing, and custom clearance.

The decision not to open a production site, at least in the initial phase of the project, in the country in which the distribution of the new tracer has been planned, and use instead the product capacity of a neighboring region, needs to take into account all the points above, and ponder the possible high economic impact, which can worse in case of negative distribution results.

In my personal experience, this approach may also have an additional negative aspect; if the customer's expectations are not met, an unfavorable sentiment can rise, and the environment become resistant to the subsequent commercial production phase in the territory in question.

All elements, therefore, must be analyzed with extreme attention.

Sharing responsibility between the corporate commercial/business development and local teams is of paramount importance. On the bases of my experience, I found a number of points that should be considered and perhaps used as a model in the initial business definition of a PET tracer and its on-going monitoring.

(a) The local team defines the sales forecasts for each CMO site: expected doses, days required for a period of at least 5 years from the theoretical start of marketing, and operative margins.

(b) The corporate analyzes and compares the forecasts with those of other countries but does not change the data.

A critical review of the corporate, carried out in advance and based only on the statistics of other countries, could invalidates, in my opinion, the initial assessment of the local team; the PET business is a result of specific aspects linked to the territory and the corporate may not have the possibility to know the individual situation in each country.

The local team can be guided and trained, e.g., on events that have occurred in other countries (direct comparison between local teams of different countries would be useful), but should not be forced to revise the estimates, upwards or downwards.

(c) The local team, in the operational phase of the CMO, has to be monitored but not with the purpose of assigning blame or applying correction, in case results are not in line with plans. The purpose of this kind of monitoring should be cooperative and supportive, e.g., through specific training on the product and its potentiality.

(d) The local team remains the only responsible for the success of the PET tracer.

(e) The corporate, in case of positive development of the project, should gradually reduce its interventions and disappear as supervisor when the local market becomes independent. The corporate role will continue to have a role in case of supranational negotiations (e.g., synergies between CMOs network, introduction of new products, R&D headlines) that require a general vision of strategies, that cannot be analyzed locally.

(f) The corporate team should remain solely responsible for the failure of the PET tracer project, except for the situation in which the commercial strategy and selling forecast by the local team were not correctly defined.

In this case, the task of the corporate, in particular of the supply chain, is to investigate and understand the deep root cause for the failure. The target remains to achieve the objectives and, should it be necessary, the times and methods of supply of the radiotracer should be revised together with the CMOs, and check whether they are in line with customers' expectations and/or new organizational situations. Insignificant change may play a key role. For instance, a reason of failure could be a wrong shipment strategy, such as missing to deliver the product to the closer customers first and then extend the distribution to farther locations. Likewise, new organizational decision could sometime take place, and new persons with new strategies review the initial plan, without the proper analyses of the state of play due to poor knowledge of this kind of product.

(g) The production availability within the CMO could be modified due to internal organizational changes; in particular when a long time, e.g., years, has elapsed between the definition of the production contracts and their actual implementation.

When the number of PET tracers increase, many actors become involved.

Over the years, new tracers can develop in different medical areas: e.g., oncology, cardiology, neurology. As mentioned above, the business development reference figures within the corporate may have a quick turn-over, while local staff, supply chain structures (production and distribution management team), and the organization of the CMOs themselves could remain substantially unchanged.

The approach to the development of the tracer, decided by specific corporate people, could therefore become different; this is a very delicate moment, as the local commercial representatives may not understand the reason for a sudden change in strategy.

In these situations, the educational and support role of the supply chain is fundamental, because it owns the historical memory of all products, and has matured a deep experience of both commercial and production dynamics in the various territories. The central coordination is important, at this point, to harmonize the different needs and expectations of the corporate development teams (marketing and commercial) for the various tracers, possibly grouped in a single team, to avoid different communication channels and consequently dispersion of the information.

2.3   Three Different Manufacturing Contracts with the CMO: For Mature, New, Investigational Tracers

Some important basic information should be considered since the beginning:

1. The PET network is in most of cases, at least in my experience, constituted by three main actors: the corporate team, that coordinates the national teams in the different subsidiaries, that operate in the territories through contracts with the CMOs.
2. In principle, the CMOs can produce three different typologies of radiotracers: mature tracers, new tracers, and investigational tracers.
3. National contracts with the CMOs are managed by the corporate and can be customized according to the local needs or rules.
4. There is a different situation whether the CMO is a stand-alone firm (the site itself) or is a part of a network (large national company or multinational organization).

A "mature" tracer is a radiotracer for which industrial production and distribution have already started in one or more countries. Opportunities and problems, such as customer needs, logistical, commercial, financial aspects, and competition with other tracers, are already well known.

The corporate negotiation with the CMOs, therefore, begins with the awareness of the involved risks. From an operational point of view, the CMOs can base its analysis on results already achieved through a highly standardized process, in terms of both production and distribution.

If the CMO is a country-based or multinational company in which some of its sites are already manufacturing the same radiotracer of the contract, all the necessary information is already present and a detailed risk-opportunity analysis for any kind of business development of the radiotracer is easily achieved. From a certain perspective, this is the easiest situation for both parties in the negotiation, due to the high quantity of data available.

A "new" tracer means a new product for a future market.

Even if the awareness of the commercial possibilities has been evaluated by corporate, these have not yet been tested in the field. The tracer, if aimed at a new diagnostic use, with respect to already distributed products, will require an additional organizational effort for the creation of connecting points between the specialty doctor (e.g., oncologist, cardiologist, neurologist) and the nuclear medicine specialist.

The CMO(s) should not have relevant hurdles in running the production, because production reliability and effectiveness are granted by a process that is described in a marketing authorization, even if not always tested on large volumes and therefore with limited production data. The CMOs will decide to be partners in the manufacturing contract and accept the financial and logistic aspects, on the basis of the credibility and quality of the data provided by the corporate.

In my opinion, the level of difficulty for the commercial management of a new tracer is medium.

An "investigational product" means a product under assessment in clinical trials. It is usually a not-for-profit product.

When the product is in advanced testing phases (phase 2 completed), within the limits of local legislation and within a time window that closes with the approval of the marketing authorization, it can also be commercially distributed, in this case through the "NPS" (nominal prescription shipment) model. The product is shipped only following a nominal patient prescription that is issued by the referring physician, who is obliged to direct supervision and responsibility of its use.

This is only apparently a simple solution, and it results in a complicated management of both the technology transfer and the production phases.

The relationships and the risk sharing between corporate and CMOs can be very different on the basis of both the category of product under consideration and on the nature of the CMO. A quick summary is reported in Table 2.1.

The most difficult part in the table is to evaluate the interest of a multinational CMO on a long-term project aimed at an investigational product; then, provided the first point is positively accepted, what level of confidence the corporate should provide to a multinational CMO in order to support its proposal.

In my opinion, it means to provide clear information on the following points:

(a) Market interest for the novel tracer
(b) Definition of the network of clinical centers, effective subject recruitment capacity, timing of the different clinical phases, and support by key opinion leaders
(c) Temporal succession of the tracer launch in the different territories

Table 2.1 Risk sharing between corporate and CMOs

PET tracer	Level of interest for a PET tracer				Level of difficulty in the negotiation		
Tipology	Level of commercial risk	Level of difficulties on the interaction between specialist doctor and nuclear medicine specialist	Level of tracer' interest for a stand-alone CMO (not part of a network)	Level of tracer' interest for a multinational CMOs company (more than one manufacturing site)	Economic negotiation	Technical negotiation	CMOs availability for production shipment in the early hours of the morning
Mature	+	+	++	+++	+++	+	+++
New	+++	++	+++	++	++	++	++
Investigational product	N/A	+–+	++	a	+	+++	+

+ = low, ++ = medium, +++ = high

[a] The level of interest is proportional to the corporate ability to share a long-term project with the multinational CMO

(d) Clear indication of the production sites in which the corporate is interested, for both clinical trials and commercial phase
(e) General indications of the number of lots required per month and shipping time per production center, in the industrial production phase

It is extremely difficult, to have all this information at the initial stages of a clinical study. However, if corporate wants to start attracting attention and possibly availability of future production slots, it is necessary to initiate this type of dialogue and keep it constantly alive.

Similar efforts have also to be applied with "stand-alone" CMO sites (not part of a PET network).

2.4 Customer Order Management: CMO or Corporate Responsibility?

The territorial distribution of the radiotracer in the PET network is fundamental during the entire life cycle of the tracer.

The responsibility for collecting orders from customers can be held by the corporate (through affiliates in each country) or by the CMOs: the choice to proceed in one direction or the other is closely linked to their organizational structure.

The direct management of orders by the CMO is possible, in my opinion, only if some prerequisites are satisfied.

1. The CMO has an internal customer service structure, which is available for a continuous interface with both customers and the corporate. This is the case, for example, of CMOs with a multinational organization or companies with an effective distribution channel in the area.
2. The territorial area that is served by one specific CMO (single or multiple manufacturing sites) cannot be served by other different CMOs, even if they are part of the same PET network, due to exclusivity or logistic reasons.
3. The customer, for placing an order, must contact the customer service of only one CMO also in case of CMOs with multiple manufacturing sites: in this latest situation their customer service function must be centralized.

Instead, direct management of orders by the CMO is not possible if the following situations are concerned.

1. The CMO does not have a customer service structure capable of collecting and processing requests, and maintaining a continuous interface with the customers and the corporate. This is the case of small manufacturing sites managed by Universities or Hospitals, which were established for "in house" local use of radiopharmaceuticals. Here, the local assumption of responsibility for the management of orders requires an organizational effort that is difficult to face and not economically convenient.

2. In the same territorial area, there are different CMOs with no connection among them, and consequently there is not an effective channel for directly managing customer orders. In this situation, the limit for the direct management of orders depends on the following points: (a) production capacity; (b) management of partial or total production failures.

When the orders are received directly by a CMO, and its production capacity is exceeded, the only possibility to satisfy all the requests is to produce a new batch and reschedule some doses therein. Some customers may receive a delayed delivery.

If the requests are managed directly by the corporate team, they can be distributed in to different CMOs operating in the same territory, based on their different production capacities and actual production plan. Furthermore, having such an information centrally, it would allow the corporate to expand the efficacy and possibility of back-up management. The possibility to quickly and efficiently organize the back-up can strongly facilitate the management of partial or total production failure, that it is not in the single CMO capability.

The entire flow of orders of markets that are not yet mature should be only managed by the Corporate. It is essential to establish both the correct commercial strategy and the profitability of the product. This activity cannot be assigned to a CMO and must be based on corporate direct business analysis.

2.5 Collaborations with Non-industrial Entities (Hospitals or Research Centers). Two Different Mindsets, How to Reconcile Them?

During my work, I was in the position to build a PET radiopharmaceutical manufacturing site inside facilities that were managed by non-industrial parties.

I like to mention one example in particular. The building in which the site would be housed was inside a very large international center, entirely dedicated to do research in various sectors, including radionuclide production (and this was the trigger of the project).

Any decisions even on safety issues (apart from radiation protection which was ruled by the national legislation) were made with considerable autonomy and delegated to the local teams of experts in the various disciplines. This appeared as a promising situation.

However, the first difficulty was in the very basic definition of processes: in particular the approach that could align the two different realities, an industry and a research entity, that would have coexisted in the project.

On the one side, there was a private company, supported by a strong staff of consultants, whose skills in managing processes were well defined and developed in previous national and international experiences. The industrial team was obviously focused on a quick conclusion of the project, in order to start as soon as possible the business and collect the fruits of the investment. On the other side, there was a milieu that, having had pure research as its priority, was approaching each step of the project in a very comprehensive way and with an "academic" team attitude, quite often losing the sight of the ultimate goal.

Since the initial stages, I realized that we had been completely misunderstanding the approach.

My position at the "kick-off" meeting can be summarized in a simple statement: the Research Center and my Company, according to the initial agreement, were expecting us to focus entirely on the project, and concentrate all resources and efforts on it.

At the end of the meeting, I was approached by a representative of the center who, politely but clearly, explained me that if I would have managed things with the same model used at a company, the project would have failed.

I had to take a step back and understand the environment where I was sent to work.

As a first consequence, the timing of the project had to be aligned with that of the center; and to this purpose, I had to understand the organization of the management group and the rules through which decisions were made. Then, the project, with the various activities, was "unpacked" and divided into parts.

The research center was served by many utilities that had to be reanalyzed and realigned to higher standards than those actually existing to address the need of our new project. This was possible only with the close support by the specific staff at the center, who was in charge for this type of activity, and had the knowledge of general services, security, IT system, gas lines, of the entire facility.

The technicians of the center were very skilled and specialized and would have worked for us; that was very good. However, they had somehow to change their way of working; likewise, we had to change our way of interacting with them. It was a sort of a "crash" between two completely different mindset approaches; they were used to a great freedom and time availability to analyze and act in case of a problem; we had very organized plans and strict time window between thinking and acting.

For example, in case of technical failure or problem, our position was for a near real-time intervention, to avoid any blocking of activities. But even during the qualification phase, we verified that speed of actions was not in their standard operation mode. This eventuality was not considered during the discussion of initial agreement.

When a critical situation emerged, my approach to its solution was on a double level: an action to fix immediately the contingent problem and also an effort to define a process, together with the center staff, that would work more efficiently in the future, and give us better guarantees in case the same problem would occur again.

In some cases, the analysis also involved changing the level of service provided on equipment, for instance, through an increase of routine maintenance activities that would have reduced the future risk of malfunction or failures.

Therefore, rather than immediately asking for a generic "maximum service level" since the initial phase of the project, I generally challenged the system limits emerging during the qualification activities. In parallel I worked on stimulating the many technicians of the center to improve their response by involving them in the qualification activities and also giving them the time to fully understand what we were doing. In the end, I achieved the result of creating a robust system, which indeed was confirmed during the industrial production.

Clearly, due to these hurdles and adjustments, the project duration was extended by several months. Looking back, with the experience I gained, I don't think I could have done more.

I was focused more on people, on their interests and availability, rather than on trying to secure a rigid application of contractual guarantees to the company.

Maybe I was just lucky, but I am aware that without people availability, confidence, and skills, the project would have miserably failed.

2.6 Monitoring of Production and Distribution Reliability in a PET Network

Production Reliability

Production reliability is one of the key elements to be kept under strict control along the life of a PET production site, whether it is a stand-alone site or a part of a network.

In the case of CMOs, the corporate role becomes an integral part of the analysis, since events like failed or partially produced batches can heavily impact the financial outcome.

However, as already mentioned, the number of key performance indicators (KPIs) to be monitored in sites with a short manufacturing history should be limited to the most important ones, and focused on the emerging operational difficulties they are facing.

In general, the understanding of operational performance requires at first a work of classification of the various events.

This is to avoid that each site, or territory in case of a PET network involving several countries, could develop a different terminology to indicate the same event, which would end into a non-homogeneous analysis. It is important to avoid that events that are not well defined and incorrectly classified enter in the statistical analysis, with the result of mislead decisions, wrong solutions, and delay or bias in the definition of an effective corrective action.

The first step from corporate is to clearly identify the delivery status of the amount of radioactivity requested by the customer (doses produced from the manufacturing site and their description). A possible classification is reported in the table below (Table 2.2), summarizing the definitions that have been recognized as common and essential, starting from an international professional experience.

The definition that most easily, by its nature, requires sub-categories is that of "Failed (production site)." In fact, it can be linked to different types of technical

Table 2.2 Delivery status and description

Dose status	Description
Canceled (customer credit)	Cancelation communicated by the customer beyond the agreed notice time (e.g., 24 h before production). In this event, the customer may have to pay the dose, even if not anymore requested (depending on the commercial agreement)
Canceled (customer)	Cancelation communicated by customer within the agreed notice time
Canceled (order entry)	Data entry mistake. If the mistake is made by the customer, a reimbursement equal to the dose value could be considered
Canceled (production site)	Production is canceled with less than 48 h notice (2 business days). (1) Production doesn't start; (2) the production site cannot deliver the dose and the customer cancels the order (without rescheduling); (3) batch is successful but the manufacturer is not able to deliver all requested doses
Delivered	Dose regularly delivered
Delivered (Commercial—out of scope)	Out of direct commercial scope (e.g., part of the batch is used in-house for internal reasons)
Delivered (for royalties only)	Production site compensate royalties to the corporate according to delivered doses
Delivered not administered	Doses are not administered by the customer for internal reasons (patient unavailability, equipment issues)
Failed (production site)	Doses not produced due to a complete or partial (low yield) batch failure. All doses are labeled as failed. New rescheduled doses are labeled pending/delivered
Pending	Scheduled doses not yet produced
Rescheduled (customer)	Doses rescheduled by the customer within the allowed time window
Rescheduled (production site)	Canceled by production site before 48 h prior production: possible no refund on the assumption, that patients can be potentially rescheduled
Test batch	Delivery status "Delivered," if the test batch is successfully produced, "Failed (production site)" if the test batch failed

problems during the manufacturing of the batch, related to the equipment, or to the raw materials used for production.

An example of tentative classification, not necessarily exhaustive, is reported in Table 2.3.

As said, the homogeneous classification of the problems that could occur allows both accurate statistical analyses and appropriate corrective actions. An example of a potential application of this approach could be a network in which all the sites are supplied by a centralized facility or warehouse. In such a case, the use of a specific batch of a defective raw material could cause transversal problems to many sites, although not all together nor necessarily at the same time, but depending on the local stocks. If each site of the network uses the same definition of the problem, it would be easier to have a quick indication on the nature and extension of the emerging issue and trace the problem back to its root cause.

Clearly, it is necessary to dedicate adequate resources to collect and update regularly the information shared by the entire network in a specific folder.

Table 2.3 Failure classification

Delivery status	Description/interpretation	Type of failure	Classification
Failed (production site)	*Cyclotron electronic failure*	Cyclotron	Hardware–Electronic
	Cyclotron leaks or errors related to fluids, gas, vacuum and transfer lines	Cyclotron	Hardware–Fluid and Gas
	Cyclotron mechanical failure (valves, pumps, syringes, targets)	Cyclotron	Hardware–Mechanical
	Cyclotron software issue (communication, corruption, connections)	Cyclotron	Hardware/Software
	High F18 quantity produced	Cyclotron	High yield
	Low F18 quantity produced	Cyclotron	Low yield
	Cyclotron operator error	Cyclotron	Operator error
	Dispensing kit	Dispensing	Consumables
	Dispensing electronic failure (automated dispensers)	Dispensing	Hardware–Electronic
	Dispensing leaks or errors related to fluids and gas	Dispensing	Hardware–Fluid and Gas
	Dispensing mechanical failures (valves, pumps)	Dispensing	Hardware–Mechanical
	Dispensing software issue (communication, corruption, connections)	Dispensing	Hardware/Software
	Dispensing operator error	Dispensing	Operator error
	Facility leaks or errors related to fluids, gas, and ventilation	Environmental monitoring system	Hardware–Fluid and Gas
	Facility mechanical failure (ventilation, pumps, valves)	Environmental monitoring system	Hardware–Mechanical
	Facility software issue (communication, corruption, connections)	Environmental monitoring system	Hardware/Software
	Environmental monitoring systems	Equipment issue	Counting system
	Facility electronic failure	Equipment issue	Hardware–Electronic
	Facility leaks or errors related to fluids, gas and ventilation	Equipment issue	Hardware–Fluid and Gas
	Facility mechanical failure (fridges, freezers, ventilation, pumps, valves)	Equipment issue	Hardware–Mechanical
	Facility software issue (communication, corruption, connections)	Equipment issue	Hardware/Software
	Synthetizer cassette/consumable not working properly	Synthetizer cassettes/consumables	Consumables
	Synthetizer cassette/consumable high performance	Synthetizer cassettes/consumables	High yield
	Synthetizer cassette/consumable low performance	Synthetizer cassettes/consumables	Low yield
	Synthetizer cassette/consumable-operator error	Synthetizer cassettes/consumables	Operator error
	Synthetizer electronic failure	Synthetizer HW/SW	Hardware–Electronic
	Synthetizer leaks or errors related to fluids, gas, and vacuum	Synthetizer HW/SW	Hardware–Fluid and Gas
	Synthetizer mechanical failures (valves, pumps, waste bottle)	Synthetizer HW/SW	Hardware–Mechanical
	Synthetizer software issue (communication, corruption. connections)	Synthetizer HW/SW	Hardware/Software
	Synthetizer high performance	Synthetizer HW/SW	High yield
	Synthetizer low performance	Synthetizer HW/SW	Low yield
	Synthetizer-operator error	Synthetizer HW/SW	Operator error
	Sterilising filter	Filter failure	Consumables
	High yield	High yield for unknown reasons	High yield
	Low yield	Low yield for unknown reasons	Low yield
	QC equipment electronic failure	QC	Hardware–Electronic
	QC equipment mechanical failure	QC	Hardware–Mechanical
	QC equipment software issue (communication. corruption, connections)	QC	Hardware/Software
	Out of specification (drug product specs not including RAC)	QC	Out of Spec

During my experience of managing a PET network, in which the reliability of the sites had to be weekly reported, I found it particularly effective to have a specific event-by-event comparison with the technical and QA support.

Sometimes, when the manufacturing reliability of the CMOs site underwent a deterioration, which can occur even in experienced and mature sites due to an excessive self-confidence, it was useful to have the corporate asking for a retraining, with the support of expert personnel, of the entire staff on the production process.

Another risk, often latent and possibly emerging suddenly, is that CMOs can modify plants and equipment over the years. The adopted changes are in most of cases communicated to the corporate, but sometime delayed notification or insufficient details do not allow a correct risk assessment, and limit the ability to foresee possible operational implications, until problems occur during the production phase.

Production Reliability in Pre-production Start-Up Phase

All the elements reported in the previous paragraph deal with the control of production reliability, that generally begins after the start of regular production activities. The question is how can we prevent the problems: competence and experience are key factors. Thus, one guiding principle in the selection of a CMO could be the presence of qualified personnel, with good experience and consolidated history in the production of radiotracers.

Another critical point is the level of maintenance of the equipment at the site; as specifically required by GMPs at which level is set and "who" performs the maintenance are two aspects which are often not analyzed with sufficient detail and technical focus.

Therefore, in my opinion, this last aspect must be carefully considered, for instance, by using the following considerations.

– Is the level of maintenance aligned with the stress to which the equipment is subjected, i.e., the number of batches produced? For each single critical piece of equipment, it would be extremely useful to ask for the support of the manufacturer, in providing available studies on the robustness of equipment components during the critical stages of operation and, whenever possible a local stock of more critical parts (difficult to receive or frequently replaced).
– Is maintenance carried out by the internal support staff, by the manufacturer of the equipment alone or by a reasonable combination of the two?

In our experience, an effective and efficient maintenance is not achieved only when it is provided by the manufacturer. Many sites reached a near 100% reliability, and maintenance was handled mostly by the internal staff.

The difference is made by the level of certified training received by the internal staff, usually provided by the suppliers themselves, and the contractual service conditions, in terms of assistance in difficult cases and speed in the replacement of spare parts.

However, it is very important to create a solid and robust interface between the production staff and the equipment supplier, in particular when an internal resource is not sufficient. And this situation, unfortunately, is very frequent.

In the case of a CMO with an already solid manufacturing experience, it would be useful to provide the corporate with reliability data, for example, of the last 2 years. Thus, the corporate would be allowed to understand the real level of technical preparation of the potential partner more easily.

Liability Shared Between Corporate and CMO

The liability boundary between corporate and CMO is not always clear-cut.

This happens, for instance, when the corporate supplies some or all of the raw materials for the production. An example could be represented by the synthesis kits (plug-in cassettes containing all the reagents) or the analytical standards for quality control.

If a problem arises from the use of these materials, a dispute between the two parties can easily take place. Often the mutual responsibility between corporate and CMO is understood only after years of production statistics or in the case of anomalous reliability peaks linked to specific batches of raw materials.

The possibility of assigning a compensation to the CMO, for instance, with a partial refund of the batch value (e.g., 50%), can be considered by the corporate if the parties agree on the evidence of defective raw materials.

Risk sharing can be agreed in the contract since the beginning, as a guarantee that both parties will do their best to solve any problem timely and efficiently, in order to continue the production agreement.

However, there are other elements that could lead to a dispute. I refer not only to the production process, which is property and responsibility of the corporate, but to other raw materials (such as primary containers or consumable kits used in product dispensing) that, although not directly supplied by the corporate, are listed in the marketing authorization.

For example, in case of a shortage of raw materials, e.g., due to a unilateral decision of the supplier, there may be the impulse to delegate the CMOs to search for a valid alternative and leave them the process of identifying the new supplier. This solution, in my opinion, is feasible only if the modification in question is limited to a single CMO.

In the event that several CMOs are involved, it may be a good practice for the corporate to actively participate in the search of a new supplier, for the following reasons.

1. The presence of the corporate allows a clearer view of the total volumes necessary to satisfy the needs of the entire network for a specific time frame, such information is unknown to each individual CMO.
2. Larger volumes allow to negotiate better prices and standardized conditions of supply.

3. A full confidence in the analysis of the specifications of raw materials, which must be aligned with those declared in the MAA. The opposite case will necessarily involve the change of the MAA and consequent costs.
4. Coordination of the qualification activities, that are performed inevitably by individual CMOs, who have to verify the performance of the raw material on their specific equipment (e.g., changing the primary containers).

Distribution Reliability

Production and distribution reliability are not always correlated and inline and there are sites with production reliability greater than 95% facing a series of distribution difficulties and, vice versa, sites with good manufacturing performances but issues with the delivery of products. In general, the negative alignment between production and distribution reliability is indicative of organizational problems.

Once again, the role of the corporate can be fundamental: assigning challenging delivery times to a CMO, such as in case of long-distance shipments by air or by road, translates into an increased risk of delays and therefore of low reliability.

Before assigning these tasks, it would be always advisable for the corporate to check the feasibility of product deliveries, according to a risk assessment.

The same routes could have a variability, linked, for example, to seasonal periods and holiday shifts. Of course, all reviews of distribution performance should include the detailed analysis of the nature of the difficulty in the deliveries.

Usually, the CMO, like the company, does not have its own carriers, but relies on the service of external companies. My suggestion is always to assign the management of the transport company directly to the CMO. During the review of distribution performances, it could be useful for the corporate to establish sessions of analysis in which the transport company is involved together with the CMO. The experience of the transport service could be important to highlight possible difficulties of the different routes and the possibility to validate alternative solution in case of critical situations.

Chapter 3
Elements to Take into Account for an EU PET Company Acquisition (AP)

Contents

The most important moment in my professional experience was when my company decided to sell the Italian PET production branch of which I was on-board.

As soon as I was officially informed of the decision, I had the opportunity to participate in all the administrative/regulatory phases, necessary to finalize the agreement with the purchasing company. I was on the side of the purchased PET company.

The first step I had to deal with was the transferring and/or closing of contracts, modifying licenses and authorizations, and managing the unavoidable "human" impact. All these aspects, in my experience, had a level of complexity, if possible, higher than the creation of a new organization.

Furthermore, since the peculiarity of the PET sector, in which both pharmaceutical and radiation protection legislations are strongly connected, my impression was that for an external company specialized in acquisitions, it was extremely difficult to understand the entire scenario of operation and distribution activity.

I have tried to summarize below the key technical points that, in my opinion, have to be taken into consideration in the practical execution of an acquisition operation.

Obviously, I was not specifically committed, except as a technical advisor, on the aspects related to the initial feasibility of the acquisition, like the due diligence, which is usually assigned to the legal department.

© The Author(s), under exclusive license to Springer Nature
Switzerland AG 2023
A. Pecorale et al., *PET Radiopharmaceutical Business*,
https://doi.org/10.1007/978-3-031-51908-6_3

3.1 Human Resources

The company obligation toward the employees can be different from country to country, based on the local laws. For example in Italy, if the business unit sold by the company is completely independent, the manufacturing department and the logistic service must be both acquired, since the two activities are strictly connected in the manufacturing process.

In some countries, there are local regulations to protect the business against external acquisition, especially if the company has received in the past government contributions. Therefore, legal and HR analysis are needed, based on the local organization (structure, number of employees, obligations) and according to the law of the country in which the company operates. This activity should be reviewed by an independent body to guarantee the impartiality of the analysis.

In my experience, it would be wise to consider also the possibility that the regulatory department (e.g., quality assurance, regulatory affairs, pharmacovigilance) of the selling company continues to provide support to the acquiring company for a reasonable period, at least till the processes are fully stabilized, according to the new set-up.

3.2 Regulatory Authorizations

(a) Radiological license and site decommissioning. It is better to carefully verify the expected timeline of the transfer of the license to the acquiring company. In Italy, for example, it could take quite a long time due to the involvement of different ministries, which have to express a concerted decision.

   The decommissioning of production and laboratory equipment requires that they must be inventoried, checked, and disposed off with the support of the radiological qualified expert (usually an external service) and with the support of specialized waste-disposal companies. This activity deserves a special care especially if the site has a stock of contaminated wastes: for example, the enriched water used in the production of the precursor radionuclide Fluorine-18 may contain tritium.

(b) GMP licenses. Regulatory Affairs must provide the acquiring company with the list of national GMP manufacturing licenses for each country, including their release and expire dates (if applicable).

   The list of deviations raised in the last three Health National Authority audits must be communicated (mandatory) in order to evaluate the possibility that the company has taken specific commitments with the National Authority in terms of QMS modifications or structural upgrade. The local timeline needed to transfer the license to the acquiring company, and if it is linked to a preliminary site inspection must be known.

An important social/ethic aspect is also to be considered: in case the acquiring company would dismiss one or more production sites, the potential lack of tracer availability in a specific territory could give raise to an obligation issued by the local Health Authority that the acquiring company continues to guarantee the production.

(c) MAA licenses. RA must provide the list of MAAs that have been granted, including mutual recognitions, the timeline of MAA approval (if applicable), and the expiry date for each country. The possibility that an existing or imminent update of compelling regulations, such as the EU pharmacopoeia, could affect the MAA must be evaluated, as well as the expected timeline for MAA transfer in the country of the acquiring company.

Once the result of the analysis of points (a), (b), and (c) is clear, a possible option is that the selling company will temporarily maintain the ownership of the licenses, leaving to the acquiring company only the finance and strategy control of the transferred branch.

3.3 Business Aspects

The financial aspects of all tenders, suppliers, third-party contracts, assets, and R&D collaborations should be considered.

1. Tenders

    The selling company must provide, per each product, a detailed list of the ongoing tenders in each country, including sales volume, delivery timelines, and their expiry date for each customer.

    In case the acquiring company would rearrange the production capacity to manufacture other radiotracers, any obligations of actual tenders will anyway remain a mandatory commitment.

    Furthermore, it must be clearly agreed in the contract whether the closure of legal disputes concerning tenders remains a responsibility of the selling company.

2. Suppliers

    A detailed list of qualified suppliers and related contracts/commitments must be presented by the selling company. In fact, the potential changes that could be adopted by the acquiring company, in terms of preferred suppliers and/or changes in product priority lists, could affect commitments already taken with some existing suppliers. Typical examples of these suppliers are providers of maintenance service, transport companies, external analysis laboratories, and raw materials dealers.

3. Third-party contracts

    A detailed list of third-party contracts must be provided by the selling company, for example, external laboratories that are performing part of the manufacturing activity.

In case some activities are performed by a branch of the selling company that is not included in the transfer deal (for example, sterility test), the possibility to negotiate, at least temporarily, the continuation of the service together with a convenient price should be evaluated.

This is because the introduction of a new external laboratory in charge of tests that have an impact on the release of the product could require the introduction of changes in the site license or in the MAA.

4. Assets

A detailed inventory of assets must be provided by the selling company.

All the costs related to the dismission of the assets at the manufacturing facilities, external suppliers, third-party facilities, research centers, that are not part of the PET acquisition will remain under the responsibility of the selling company, unless otherwise specifically agreed.

5. R&D Collaborations (if applicable)

The selling company must provide the existing list of collaboration with research centers (pure R&D activities) and ongoing clinical studies, so that the development of product with social interest is not interrupted and the business profitability preserved.

Chapter 4
An Old Laboratory, in the Cyclotron Building Front Lake, and Its Strange Inhabitants (AP)

It was not the first time I went to visit a PET production site, not in that area surrounding the lake.

The request of a site inspection came on a Friday afternoon, I was already closing my week of work, as usual by making an overview of the next week plan. In some "time management" company courses, they said that this operation can reduce the stress: I didn't believe it much, but I kept doing it.

The radiopharmaceutical site in which I was working seemed to have lost its good performance: a too high number of production failures linked to human errors; the inspection to a new possible PET manufacturing site could be a good opportunity to mitigate my worries.

The visit was scheduled for the following Tuesday, I got ready to reschedule the activities, sent a few e-mails and stopped thinking to the working issues, at least for the weekend.

Tuesday morning, I left my home. My destination was not far, a couple of hours by car, the weather was terrible, November was particularly rainy; windshield was so much wet that made the brushes difficult to move.

A. Pecorale et al., *PET Radiopharmaceutical Business*,
https://doi.org/10.1007/978-3-031-51908-6_4

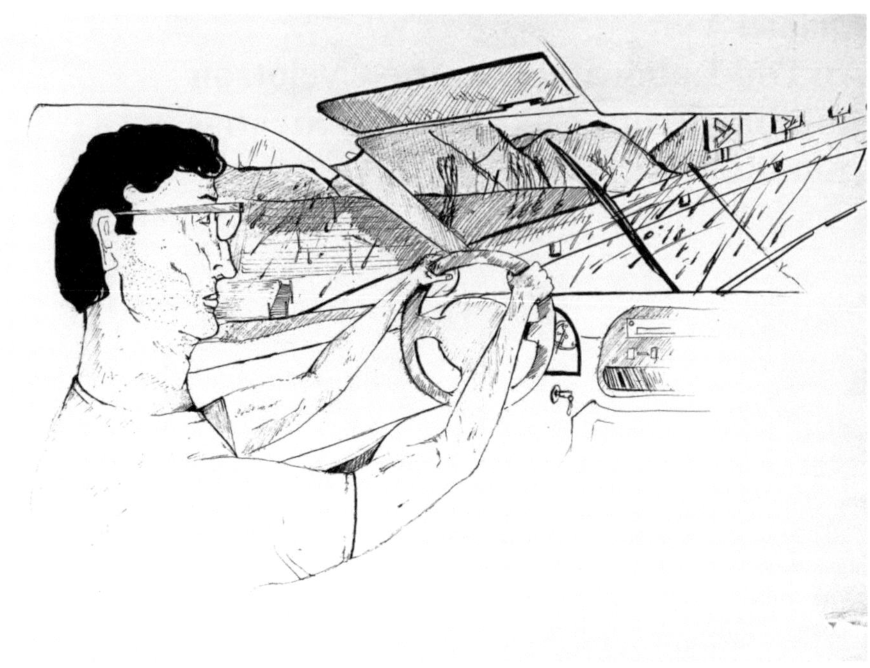

I took the road alongside the lake: deserted villages and no one along the lakefront. Arriving nearby, I struggled a lot to find the address of the cyclotron building, the place was isolated.

I left the car; I have never enjoyed driving and the context surrounding was not the best one.

Entering the site, welcomed by the reference person, got rid of my soaked overcoat and, after the introductions, we had a coffee together, in a break room having as only furniture a red vintage sofa.

I was expecting to face the usual articulate and eloquent attitude by a scientific person working in a research center, instead I was facing a very kind but inexplicably embarrassed man; I could perceive this in his agitate speaking and moving.

My feeling raised when I asked him to visit the plant and speak with the operators. He suggested, instead, a detailed presentation by him in which he would show me all the pictures of the environments, no production was in progress at that moment and also all the operators would have left the site shortly: it would not have been useful to insist on this kind of program.

Instead, I insisted and I won.

The PET laboratories were inside a large building, not concentrated but distributed in several areas.

I entered the controlled area, indeed there were very few people.

I recognized and greeted a technician I had already met on other occasions.

The production area was equipped with qualified instrument, at least so it was claimed in the maintenance schedules hang on the walls.

The quality control environment was a bit small, but everything was in order, even the register of use of analytical standards, which normally are neglected.

Entering the warehouse, adjacent to the quality control and connected to the production rooms and the corridor through a material passage, I was struck by the extremely limited space, just a couple of square meters.

I asked to open the cupboards and to see the raw materials inside. Confined in their well-defined and identified spaces, the quarantine material was separated from the approved one: there was also an aluminum case with padlock, containing any rejected material, an effective solution.

I went into the corridor together with the reference person to view the cyclotron control room.

It was an old Scanditronix from the early 80s, an old-fashioned instrument compared to modern equipment. The room was impressive, with a huge control desk.

The only note of modernity was the flat screen of the control computer, otherwise everything suggested a spaceship from the "Apollo project," with only analogic instrumentation. But, judging by the key indicators, it seemed working quite well yet. Of course, weekly maintenance every quarter was perhaps a bit excessive, but with such complex and old equipment, it is better to abound in some precautions.

As I bent over a button covered with tape and with a metal lid, I stopped. A thought about my production project crossed my mind. I turned to the reference person and asked him to confirm if a couple of productions a day were in the capacity of the site. He confirmed, but I noticed that he was agitated.

I then asked, and I saw uneasiness in his eyes, how was it possible to have so few raw materials stored in the warehouse or was there a second and larger storage area?

He said there was a second storage point in the basement, but it was not an usual standard warehouse, officially reported, rather a room for bulky materials and packaging, which did not enter the production cycle. I asked to see it, but I had to insist a lot also in this case.

The manager would have been willing to show me the registers of the material and pictures, the place was difficult to reach. I had to insist again and again.

The place could be reached in two ways: the first one through a concrete staircase, which descended from the local mechanical laboratory, passing through an old rusty equipment and a corridor, the second one directly reachable from the outside, through a ramp for vehicles.

We decided for the second option, being the first option more dedicated to maintenance activities. In this way, we also would have been able to inspect the entry and exit flow of materials.

The entrance to the basement was dark, a few dim neon lights illuminated the environment. From this we entered a low room, the ceiling was crossed by the pipes of facilities.

Cupboards and room were crammed with documents and the spare parts of old equipment.

But in front of us, was well visible the signal of the Company logo and the indication of the "warehouse."

When we were walking through the room, one of the neon went out. I turned and saw the reference person standing beyond the cupboards, hesitating. I invited him to open the door, but instead he repeated again that it was not necessary to inspect this warehouse: the place was not well organized and we were just wasting time.

He added, which surprised me, that it was nearly 4 p.m. and that soon we would be alone, except for the security guards who periodically inspected the buildings. I insisted, in a conciliatory tone, and he reluctantly opened the door.

The content was in line with what was previously indicated by him about packaging material items, but the warehouse, also contained some production accessory materials.

In the description of the plant, there was no mention of this storage point so, I claimed, it was necessary to make it visible.

I understood that the inclusion of this warehouse, so organized, would create many complications. For example, I imagined the difficulty of reorganizing the entrances by removing all things that obstructed the access, as well as defining a regular cleaning process and correct identification of the environments. But all this had to be done.

I left the warehouse heading for the exit; the neon, which was running out, was pulsing.

Once on the ramp, I asked if a night production was planned, it was.

It was now five o'clock, dark and it was raining again.

The reference person confirmed to me that the shift operators would be in the site by 10 p.m., so I left the plant, heading to a nearby hotel.

The hotel was right in front of the lake, completely empty given the low season.

The rooms were badly furnished, only the balcony offered a magnificent glimpse of the lake, with the distant lights of the ferry boat that shuttled between the two shores.

I ate something, and at 10 p.m. I was on time at the entrance to the plant.

To my surprise, the same reference person welcomed me again, we obviously stopped for a coffee.

The corridors were dark, no noise, we headed for the controlled area, where a sleepy employee subjected us to the customary recordings.

Once inside, I turned left to meet the cyclotron operator. He had clearly been informed of my visit; everything had been tidied up.

I asked what he knew about the warehouse in the basement, if there were any materials used for the cyclotron's activities: he was surprised by the question and told me that he used to go there, only during the day, to store the production documents or, in general, for transporting materials. Suddenly the operator turned and with an excited voice said that he had to hurry to prepare the beam to produce the radioisotope, there was the risk of delay the PET production.

I left him to his activities and, together with the reference person, we headed toward the production area.

The corridor walls were made of concrete, the doors of the radiation chambers opened on the left, about two meters thick.

The first room was dark, only the steel of some metal component of the experimental targets shone. A violent blow startled us both.

A quality control operator was knocking the window that separated the laboratory from the corridor, asking the reference person to come in to view some report.

The reference person, apologizing, entered the laboratory, I remained outside.

The silence was almost absolute, except for some muffled noise that came from the production area. I approached the door to the irradiation room.

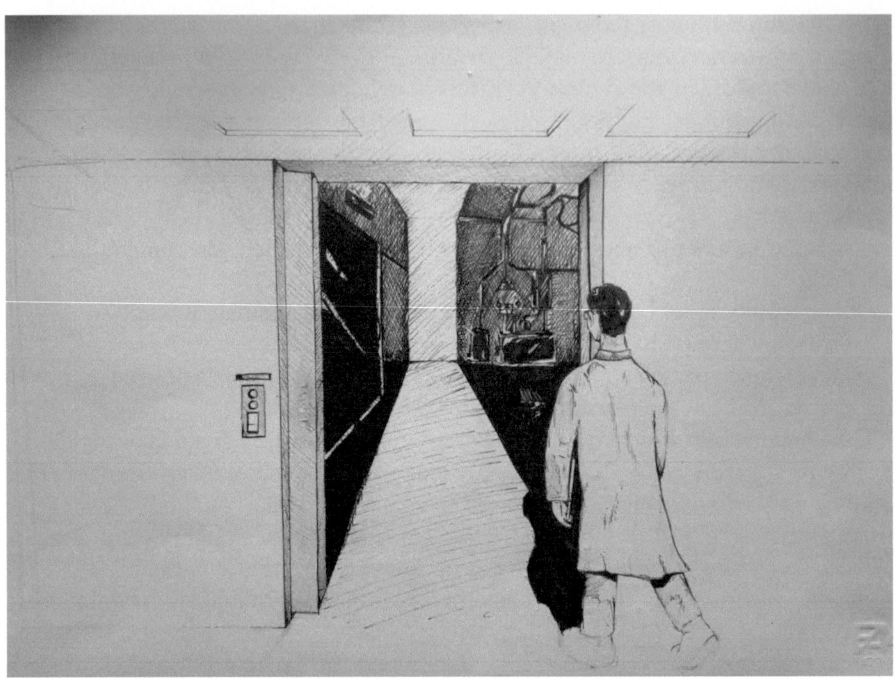

A clicking sound seemed to come from the back of the dark room, like someone running across the floor. It didn't seem like it came from some stand-by equipment, in fact everything seemed turned off.

Suddenly I heard a light laugh. I felt afraid.

I was about, ignoring the prescriptions, to turn on the light, when I felt a hand on my shoulder. He was the reference person who, after completing his supervision in the laboratory, was asking me to follow him into production area.

He looked at me and we recognized in both of us the same feeling, something similar the embarrassing we had in our first meeting.

I asked him if other activities were carried out in the area at night, in addition to the PET production.

The answer was that in the past years before the development of PET, the cyclotron was used for a couple of nights a week to produce another isotope, but now all research activities were concentrated exclusively on the normal working day.

While he was talking to me, almost running, he was heading toward the production area.

The complicated dressing procedures made me forget the strange laugh I have heard; I went into production and supervised the preparation of the synthesis module and the transfer of the radioisotope from the cyclotron.

Everything was done with great professionality; I did not see any gap in the process.

After about 2 h, the synthesis started, an automatic process, but worth to kept under control.

I was still thinking again about the laugh: was it real, or was it my imagination?

I went out into the corridor. The radiation room was dark and silent. Leaving, after the check of contamination, I went to the break room.

At the end of the corridor, I could hear the clicking of the keyboard from the office of the reference person, at least I thought it was him.

I had the coffee alone and thought that working in those conditions of isolation could be really difficult. I took my bag and decided to turn on the computer to look my e-mails: we had soon to start our project to construct the new PET site in this building.

Suddenly, however, I decided to go out, the wind had risen. I lit a cigarette and started walking around the building, founding myself in front of the entrance ramp to the basement at the end.

I saw the entrance door, slightly open, I moved down the ramp and entered; the hum of the systems could be heard on my left, the entire area, bounded by a metal grid, bordered a storage room that contain spare parts for the cyclotron. On the right, a long corridor cluttered with building materials.

I went back and entered to inspect the warehouse: I wondered how long it would take to make the environment acceptable as part of a manufacturing facility.

The neon, running out in the afternoon, was now working. Strange, have it been replaced in few hours?

I moved forward to the warehouse door. The reference person hadn't taken the key with him, in fact it was in the lock. I thought it was not correct that the access was not under control.

The key was partially rusted: probably both because no ventilation system was in place and, due to the proximity of the lake, the humidity penetrated in the room. I pushed open the warehouse door.

The boxes were placed carefully and marked with labels; no dedicated lights were present in the warehouse, but they were part of the central light system. Suddenly the door slammed shut. Strange, there was no automatic return system.

I heard something rolling behind the door, bouncing on the floor. I decided to go back and opened the door.

It was a small red rubber ball, the one of those used in the 70s, with childish images. I took it, there was no one but me.

I imagined a joke from the operators, but the activities in upstairs were so stringent that they could not have abandoned them, without risking production delays.

Further, it seemed to me that it was impossible that the operators would have been joking with me, because my visit had an official connotation. I was there as a representative of the company and the purpose of my visit was to verify if there were the condition for a collaboration that would be beneficial for both parties. So, this hypothesis was excluded.

At a certain point, I heard again the laugh, the same laugh I have heard before upstairs. What was it?

I reached the end of the corridor in the left and found myself in a room without exit. The corridor was completely dark, and I used my phone to light.

I realized the absurdity of the situation and tried to keep calm, even though my heart was beating hard.

I tried to think about rational thoughts: the synthesis was about to end, and I had to go back to the PET laboratory.

I ran my way back to the ramp to reach the entrance of the controlled area.

As soon as the operator saw me, I perceived that he was aware of my sensations or rather fears, but he said only that the dispensing process was about to begin.

After production, I asked to the reference person and staff to have a meeting for a discussion with me about possible modification to be agreed.

We went to the meeting room: it was about five o'clock in the morning and it was raining again.

They listened to my report without any comment.

In the end, I asked if, by chance, they had been listening some strange noises, like a laugh.

They said no and added that certainly the isolated environment in a so old structure could lead to strange suggestions. Some noises could have been due to the building settling or maybe some acceleration in the ventilation system.

I was not sure about their answers even if they were reasonable. They seemed not surprised about my doubts, as if they had in the past faced these kinds of requests and have ready their answer. My feeling, however, was as if there was an unspoken agreement between them, I was unable to decrypt.

So, I passed to some technical suggestions: the auxiliary warehouse could have been placed in one of the empty rooms, that I had seen near the premises of the PET facility. This solution would have greatly simplified the qualification activities.

They agreed and asked me to write it down in the report, so that they could officially solicit the management to grant the use of those premises for production activities.

I greeted them, I took my car and went to the hotel to rest.

My visit was completed in this way, the project of collaboration went on: the first PET site was constructed just in this old cyclotron building.

How to balance the rationality with something as irrational I had perceived that night?

As I could experiment during my working activities there in the following years, the same strange atmosphere I perceived in my first visit will permeate the new PET laboratory as well as all new operators.

Chapter 5
A Sustainable PET Quality Assurance (QA)

Contents

5.1 The Reasons for a Choice: Why Working in Quality Assurance?

How Did You Decided to Work in QA and What Were the Difficulties You Encountered?

My professional experience in Quality Assurance (QA) began at the end of the 80s. It started in the sector of in vitro diagnostic kits, mostly used at hospitals for the detection and quantitation of biological markers. It was by hazard, without a specific decision or selection, as often happened in life at that time, and perhaps still today.

Unfortunately, the training of new employees was not as developed as it is now, and in the new environment, I had to take my new position without a complete and adequate training, only theoretical instruction on GMP and references to relevant documents and sources of regulations. I missed at the time any education on what a Quality System (QS) was or should have been. It was not due to a lack of willingness from the company, more simply it was because the operational definition of a QS was still in its beginning; it was not very extended, nor widely known at that time.

A. Pecorale et al., *PET Radiopharmaceutical Business*,
https://doi.org/10.1007/978-3-031-51908-6_5

However, I was not discouraged by the apparently intricate matter I had in front: it was my first job in an industry, and everything was new and challenging to my eyes.

I remember, with fondness, the days spent with a colleague who, like me, was a new resource recruited to work on quality assurance. We worked intensely in the documentation service department, reviewing piles of folders full of production batch records.

We had been told that, through this type of an exercise, we would understand both the production and the quality control processes. At least in that specific situation and time, the QA consisted almost exclusively of the documentation service, and it was not more than an archive of batch records, with little contacts and almost no information sharing with any other function in the company.

As a matter of fact, we set our goal through the analysis of this kind of documentation to understand the sequence of operations throughout the entire production process, and identify the interconnections between the various functions involved: from material storage, downwards to production, quality control, and distribution of the finished product.

It was just through the understanding of the flow of operations, information, and activities among the various functions that we highlighted all connections and contact points and became able to build a map of the processes and rationalize their functioning. We did all this in full autonomy, and without guidance.

Looking backwards to this work, I still wonder how we could manage such a situation of global information confusion and poor communication with the other functions. However, it was thrilling and motivating for us, young workers in QA, to ask the team supervisors of production, quality control, warehouse, and distribution for explanations on the process they were following and manage including failures, problems, and bottlenecks. We learnt a lot from interviewing these functions, but this kind of an approach had soon positive outcome even for them, because they started to gain awareness of their importance inside the process and stopped to consider their departments as isolated islands.

We faced a number of difficulties, including a sort of resistance to communication at the beginning. The manager of a step in the global process, for instance, production or quality control, was often an expert, who had been working in that sector for many years, and considered his decision granted and obviously justified, often even too much. Under such circumstances, critical thinking often tends to flatten out and, at the same time, even a certain vein of intolerance tends to arise, so that requests of clarification were perceived as an external intrusion into one's own territory and work.

Obviously, each sector worked according to regulations but coordination and surveillance on a well-defined global framework were missing.

After months of lonely work, performed in autonomy and without any coordination, we finally got supported by a manager, coming from a pharmaceutical company, who was an expert of GMP and QA.

With her help and expertise, we could take the next step: the construction of a structured quality system inside the company. An organization that was no longer just a "Documentation Service," but an efficient system that was capable of supervising every aspect of the product production cycle, from purchasing of raw materials to distribution of the finished product.

However, facing the initial difficulties had forged a mental attitude in our minds. We had learnt how to "investigate" the actual context and describe it with schemes and flow charts highlighting relationships, sequences of activities, and control points. By this approach, we were able to easily understand and verify the actual application of regulatory standards in each activity and provide an "at glance" picture even of many complex processes, otherwise complicate to be described in their entirety.

In 1992, my meeting with the radiopharmaceutical sector was again a new shock. These products had just been classified as drugs, following an European Directive in 1989, and all Member States had to become compliant with the new regulation (i.e., having a marketing authorization approved) despite a production history of about three decades. Specifically, these products (Mo-99/Tc-99m generators, compounds labeled with various radioisotopes, such as I-131, I-125, I-123) were used as diagnostics in Nuclear Medicine for Single Photon Emission Computed Tomography (SPECT) scans. Unfortunately, at a that time, the GMP regulations were difficult to apply in the departments that produced these "drugs"; the working areas and technologies had been developed long before; even the lead shielded containers, used as a part of the packaging materials, were challenging to manage due to untraced cleaning and storage.

An abridged procedure was established in the EU Directive and applied via national laws by the member states to facilitate radiopharmaceutical manufacturers' alignment to the drug regulations. The company reaction was in a development phase according to the new legislation. Although there was a "Master Formula," a clear production recipe and the raw materials were properly handled, their reconciliation and approval, following an analysis of the process, were often not adequately and promptly documented. Even if clinical trial data could be omitted (after 30 years of clinical use!) and other simplification were allowed, a full description of the process was to be submitted to the Health Authority, and many process and controls had to be analyzed, and sometimes updated or upgraded.

This time, I arrived strengthened by my previous experience of in vitro diagnostic kits, and I had a method to apply. Starting from interviews with the personnel (again, not without a certain irritation by the interviewees), we managed to define the sequence of activity, which was the necessary basis for the traceability of operations and construct the process flow charts. Finally, we included the same processes in a fully operational quality system.

In the early years of 2000s, the QA team was involved in a new project: the development of a QS in the production of PET tracer, for a total of five new plants planned in the following years.

Our task should have been to set up a robust and reliable QS in each of the new sites, whether the site be managed by the corporate QA team directly or by an independent organization (CMO). In the latter case, the CMO site would be indirectly supervised by us on the basis of terms defined in specific contracts (Quality Agreement, QAA).

In the collaboration with a CMO, our goal was to provide support, coaching and training to the personnel of the external organization, usually unaware of GMP dynamics and regulations.

Our team was made of figures with various expertise: production, quality control, and quality assurance and worked together with the CMO staff since the construction of the new facilities. In fact, the QA teams of both parties always cooperated and supported the operational functions (production and control) since the early stages of site design, building and operation: activation and validation of equipment and environmental monitoring systems, working space qualification, and selection of suppliers of raw materials and services.

In this way, quality requirements and microbiological safety standards had been applied early in the planning phase, when the site was still setting up.

A strong effort and care were dedicated to the preparation of the very basic procedures:

1. Use of production and control equipment
2. Environmental monitoring
3. Production and control methods
4. Behavior of operators inside the departments.

Of course, this was done in strict collaboration between the personnel of the relevant departments and the QA at the CMO.

There was a contextual training and education scope, in ensuring the correct use of operating procedure at the departments, and in the development of the correct mindset within QA for the application of regulatory, quality, and safety requirements.

The setting up of the Quality System (QS) in these newly built PET sites has been considerably linear, almost "agreed," because the Corporate QA has been a fundamental point of reference from the very beginning for all the functions at the site.

Consequently, the surveillance of the local system (management of deviations, change control, and risk analysis) was considered as integral part of the work of all the functions (production, controls, QA) that had initially contributed to its construction.

What Are the Key Mandatory Points of a Quality System?

The QA team works always in synergy with all the other departments of the company.

The essential characteristics for anyone who works in this sector divide in technical and behavioral aspects, which are useful in having efficient and effective human relations and communication at all levels in the organization. I would deem few points as "key," of which I am trying to make a brief description below.

1. Knowledge of legislation. Knowledge of the cGMP is certainly a key point among the technical aspects. It is the regulatory basis of the pharmaceutical world, its application allows the reliable and traceable production of suitable, high-quality, and safe drugs for patients, but cGMP also dictates a wide set of general "how-to-do" concepts without of which it would be impossible to reach a level of operational condition that can be approved by the Health Authority.

   Even though not mandatory, the knowledge of ISO 9001 regulations is equally important, in my opinion, and usefully applicable to the sector. In my experience, this voluntary standard proved to be of great support in the building and maintenance of an integrated QS, extended to the entire life cycle of the product. In fact, the ISO 9001 standard allows the supervision and monitoring of the quality not only of the operational and control phases of the process, but also of the overall quality of service, such as distribution, suppliers (including technical and logistic services), product surveillance with the management of complaints and even commercial activities, with the aim of continuous improvement and customer satisfaction.

2. Communication skills and training. It is the responsibility of the QA team to provide training on the applicable regulations to all the personnel working in the company. Here, training is not only intended as the mere circulation of information and knowledge, but must include the direct involvement of personnel. Making the QS working as a mechanism that becomes automatic and self-sufficient, in terms of self-monitoring, should be the true scope of training. The QA staff should promote the continuous exchange of information among all sectors at all levels through a constant dialogue. It is physiologic that departments such as production will find "to do" as their first priority, in terms of completing and delivering the production batches through their actions. The QA team, should always have a "to trace" connected to doing actions and this is the main technicality that QA must develop, through continuous training, and transfer the concept of how much this is important to other functions and staff. The reliability and transmission of information, which is vital for the functioning of a system, are guaranteed by means of the documental tracking of activities. Tracing the operational activities of a process, being it production, control, or distribution, means to facilitate the reanalysis of any step, if necessary. If "doing" and "tracking" do not go together, the activity and transparency of the QS and consequently of the company are compromised.

3. Alignment and balancing. The objective of QA is to align company standards with regulatory requirements, considering both internal and external context, as well as the dimension of the company itself.

   This balance is only possible if the QA involvement occurs since the initial design stages of each activity that the company intends to undertake, and should be covered by the QS.

   A commensurate solution with business needs must always be identified through a multidisciplinary discussion, promoted, and coordinated by QA among all figures involved in any project. Any decision must obviously be compatible with the real financial possibilities and the technical capability of the company and the manufacturing site.

4. Development of awareness. The quality system includes a series of surveillance and monitoring activities, aimed at ensuring that the production processes and their resulting products are in compliance with defined requirements of quality, safety, and efficacy.

   To achieve these, all parties must guarantee a continuous passage of information.

   Surveillance activities, to be effective, require all actors (QA/Production/QC) to practice the active and continuous application of specific mechanisms, such as the recording of operations and deviations, and the definition of corrective actions.

   Quality assurance, besides the name of a department, is first of all a mental attitude that, within an organization, people develop not only following an adequate training, but through the application of its principles. This allows experiencing both the advantages and the opportunities of improvement arising from them.

   The need of maintaining highly standardized processes, by tracking and recording deviations, and finding the root cause for any of them, can be perceived by someone as an additional stress of activity and a waste of time. In the real life, it would be a very risky situation for the company.

   I had many opportunities to verify, even by the direct experience of colleagues, the importance of a robust system in specific situations, especially in the technical area.

   (a) Abnormal events in the production process. Having kept track of any activity carried out during production, including raw materials used and analytical results, makes it possible the comparison with historical records of productions, even from previous years, and, in many cases, represented a valid support in the identification of the root causes and the analysis of deviations when they occurred.

   (b) Customer complaints. Complaints are a stress test not only for the production area; in fact, can impact any sector of the company.

   Despite the negative feeling that a complaint can raise, even functions very far from production are well impressed when the system reacts in a robust and efficient way. Thus, anyone will certainly appreciate the speed and the consistency of data with which the analysis of a customer's complaint can be done in a tracked process. Though this cannot be always taken for granted! Anyway, the prompt and appropriate reaction does have a positive impact on the image of company. In fact, being able to respond to the customer's complaint in a precise, clear, and punctual manner is a demonstration of commitment and professionalism, regardless of whether the complaint is confirmed as justified.

   (c) Inspections by Regulatory Bodies. Never more than during these occasions, the importance of a QS robust is evident to the team and to the entire company too.

The lower the number of observations received during an inspection by the competent authority, the quicker will be the return to operations, and the higher the reputation of the site. Of course, the company will be prone to investments, upgrades or even new activities, by granting a rapid start-up.

A very positive outcome (a few or no major deviations) for a site at its first inspection can be a strong motivation and self-trust for the team and also a very good presentation of the project at the top management.

Obviously, this positively adorned scenario requires a forward-looking attitude by the top management such as, for instance, investing in resources for the quality system development.

What Does a "Living" Quality System Mean? How Could the Concept of a QS Be Extended to Functions Different from Technical Ones?

By "alive" QS I would mean a system in which control, monitoring and supervision mechanisms, work efficiently and in real time. It should be an almost perfect geared system, managed with the collaboration, at all levels, of any component of the organization. A QS structured in this way will tend to develop not only in departments, like production and control that work in accordance with GMP, but it will extend also to areas of the company that are not normally covered by standard operating procedures.

The concepts of QS are easily applicable in technical fields, such as production, control, distribution, already accustomed to follow the stringent GMP regulations. However, in modern companies, these concepts permeate all activities much more than in the past, including not only the production sector but also all related services (like distribution, external analysis).

Nowadays, it is difficult to think that a service is composed of independent segments: the logic of the "freelance professional," in my view, is no longer applicable.

Suffice it to say that both GMP and ISO, in fact, cover all sectors, even those historically farthest from standardization, such as marketing and commercial service.

In fact, it would be illogical that processes, already subject to strict regulations, such as the promotion of pharmaceutical products, the distribution of free samples and the organization of congresses—just to name a few—are not included within the frame of the QS procedural system.

Therefore, despite a mentality originally distant from standardization, the evidence of the benefits from dealing with controlled and interconnected processes directed the professionals in those sectors to share the advantages of an organized QS.

Once again, the first impact of the introduction of a standardization mechanism, likewise technical ones, is not necessarily simple and often may require a lot of effort also in these sectors; but when the full integration throughout the service delivered is achieved, extremely positive and virtuous synergies can emerge.

The integration of specialized commercial areas in the general QS, by the definition of dedicated operating procedures, has led, in our experience, to their active involvement within the system and has increased also the interaction with all other technical and service functions. Finally, the commercial personnel realized that a more organized system could be useful even in their activities, without hindering their freedom of decision in the commercial strategy to follow.

Sometimes it was even necessary to contain their "enthusiasm"; for example, by limiting the inclusion of redundant control points in a process, or the introduction of an excess of countersignatures, that would have slowed down operations without any real benefit.

Indeed, the prevalence of mere bureaucratic aspects over the search for a rationale but fluid process, may be a temptation; in this case, it is essential the mediation and the communication skills of the QA group to support a maturation process of all personnel, including those in the commercial area.

In this regard, I would mention our experience with the commercial and medical affair departments, in the development of a standard process for the distribution of free samples at controlled temperatures.

The difficulties in the preparation of specific procedures were connected to the extreme mutability of the market and the changes of customer's needs, which are constantly reflected in the activity of commercial colleagues.

To collaborate with them, in terms of QS requirements, means to understand their specific needs and build standard processes which, much more than in the production area, should be reduced to the essentials, but equally kept effective and in compliance with regulations. In defining these processes and procedures, the QA must also consider the professional figures involved and their specific skills and responsibilities must always be aligned with the real possibility of commercial people to be in charge for their part in a process based on ongoing situations, and compatible with their offsite work.

The distribution of free samples at controlled temperatures is guaranteed by the maintenance of the cold chain. In this case, it was necessary to make use of specialized external services (storage and transport) to ensure full product safety and compliance to regulation. Scientific sales representative freed by this task have been assigned to other duties, like tracking the request coming from doctors and transferring the information on to the distribution staff. All these activities were recorded in a detailed and dedicated database, obviously managed through a specific QS procedure.

The process is quite easy to describe, nevertheless it required months to be fully developed; in the end, it was accepted with great satisfaction by our sales colleagues, who became ready for future challenges to take part in an organized QS.

5.2   QA Team of a PET Site: How to Do a Selection for the Recruitment of Operational and Managerial Personnel

What Characteristics Should Be Considered When Selecting an Operational Figure in a PET QA Team?

A specialized scientific profile can facilitate the understanding of technological, chemical, and microbiological processes that the QA personnel will have to face; in my opinion, an university master's degree in biology, pharmaceutical chemistry, pharmacy, or chemistry would give the best results.

However, the importance of personal characteristics should not be ignored: patience, a good dose of humility and attention to details are key features in this kind of job. The QA team, in fact, will find itself operating in the most disparate situations, having to keep (or bring back) under control different processes and support colleagues in updating procedures and forms, not always falling within the team's specific background. Therefore, they will necessarily work alongside different functions, and personnel with different specializations and skills.

The need to find effective and quick solutions to old or new problems, which is quite frequent especially with radioactive products, requires the ability of listening, investigating, synthesizing, and organizing the information in a short time. The possible heterogeneous environment in which the team might operate will require also the development of good mediation skills. Furthermore, an attitude to constructive discussions is essential for the performance of monitoring activities, such as self-inspections and audit to suppliers as well.

How to prepare the recruitment interview? The first selection should be based on an extremely detailed job description; the candidates must exactly know what activities they will be carrying out.

The aspect that must be immediately clarified, because it is typical of PET sites with normally little staff, is the availability to interchange roles. Each member of the QA team will not have only a specific role, but shall take care also of other activities of the QS management: from the maintenance of the documentation system to the performance of auditing activities.

Such a situation presents both advantages and "disadvantages". it will be an opportunity for the acquisition of an extensive knowledge of all aspects of a QS, but at the same time it will require a full commitment. Companies with a larger staff than a PET site have many employees working in the QA department, and each of them deals with only one specific area, like complaints, deviations, inspections, or qualifications and validations. Surely, this is a more comfortable position, but it is not equally useful for a complete professional growth.

Concerning the selection of candidates, I have frequently considered those who already had the opportunity to work within a quality system. This means having already acquired practice with production and control processes, their "interconnections" and their necessary "functional harmonization."

There are also other advantages, in selecting people with experience in QA field:

– Knowledge of the job and confidence with the specific terminology of the sector
– Shorter times for training induction
– Application of "best practices," already learned in previous experiences
– Ability to interface with senior people

Anyway, the company must guarantee to these candidates:

– Training in the specific context and processes of PET radiopharmaceuticals
– Motivation for a strong connection with the company, just because having to deal with such a specialized and demanding production.

This certainly means to foresee a structured time for theoretical training, but above all for a practical training, through the immediate inclusion in operational processes.

In my experience as QA manager dealing with radiopharmaceuticals, for various reasons (organizational situations in radiologically classified environments), I sometimes included resources in the QA team who were previously employed in other technical departments. These solutions, if well organized, can be an opportunity to improve the service and skills of the QA team itself: in fact, their in-depth knowledge of operations can be used to better understanding the need of a production department, and drive it toward continuous improvement.

What Attitudes Should Characterize a QA Manager?

I would summarize the most important characteristics of a QA manager as listed below:

(a) Independence, serenity, and objectivity of analysis. The QA activity starts from the analysis and understanding of the processes and their interrelations; these activities must be completely independent from the specific production or commercial objectives, and at the same time be supportive. To further clarify this point: sometimes a QA analysis could result in a discrepancy with the company objectives, still putting the company in the situation of having to consider them anyhow. As already mentioned, any problem can be overcome if evaluated in collaboration between QA and the other competent functions. Instead, the "production at any cost" cannot be accepted, without considering GMP implication and, in some cases, limits and restrictions.

(b) Authority. The QA manager should have sufficient esteem, trust, credit, and authority by the company to allow, in accordance with the legislation, the timely application of procedures within the various departments. Top management must support the work of the QA manager, who should not be subject to economic and production pressures, but act as a stimulus for the alignment between regulatory and company standards.

(c) Humility and consistency. The QA manager must listen to needs at any level and talk to all functions, without "being imposing and unreasonably rigid." Imposing procedures with a merely top-down mechanism is a serious mistake: the consequences would be that procedures defined in this way will risk to be inconsistent with the actual processes and miss the cooperative attitude of personnel; consequently, they will fail sorting the expected effect or even being applied.

(d) Presence. The QA manager must take into account the emotional pressure and the strict timing of PET production. At the same time, he/she must require that each employee is aware of the need to trace every single phase of the process. The compilation of the worksheet should be considered by the operator an integral and unavoidable part of the operational process. This is the only way to achieve a repeatable production and, consequently, a safe product. Personnel training and a continuous presence of QA in the departments are required to achieve an effective result and capturing any positive and negative feedback at all levels. It is a time consuming and difficult effort, and sometime, skeptical, or loose responses challenge the commitment of the QA manager.

(e) Inclusivity and collaboration. Any regulatory aspect must be analyzed and discussed in extended teams, in which the QA manager should be the reference person. It is necessary the wider collaboration, from production to commercial functions, to the definition of procedures having shared references. Then, they must be disseminated and controlled for their application and implementation. Of course, these procedures should be revised and modified as needed, but always in an organized and inclusive way. Collaboration is not only a necessity of internal requirements and relationships. For example, the development of a track aimed at full, mutual knowledge between the manufacturing site and the suppliers is fundamental. This activity comprises many different aspects: the agreement of product and service aspects, the object of the supply and extends to surveillance and open discussions, in case of any changes and complaints.

The Role of the QA Manager: Best Practice Examples

In development of the PET facility that I followed as my first experience, the QS construction was assigned to external specialized companies. They provided both qualification and planning of activities limiting the involvement of the internal QA.

The outcome was the adoption of a very broad but extremely heavy QS, which was impossible to apply efficiently during routine operations; unfortunately, the limits of this approach became visible only after production activities began.

In fact, the small staff of the PET site was unable to keep in line all the requirements of recording and traceability prescribed by an oversized number of procedures and modules.

It was quickly decided to remodulate the system according to the effective needs, in line with the size and the capabilities of the site.

The approach started from the critical review of the processes in detail, keeping in mind objectivity and inclusiveness. The entire site personnel, at all levels, was involved in the process, under the coordination of the internal QA manager. The first step toward a more rationale system was to analyze (and limit) the body of procedures and registration forms that had included into the documentation system. By comparing the procedures in place with the actual process performed, the huge pool of documents and procedures was revised. The recording forms (and procedures) were rewritten in order to contain the necessary information that was sufficient to guarantee the traceability of the operations, or discarded when useless, redundant, without any interconnections among them.

In the end, the number of forms was reduced by 50%.

Another simplification to the QS under scrutiny was the elimination of redundant double checks in some procedures, which had been defined without any true assessment of their actual need, or the presence of an inherent risk.

During this revision, we discovered that the overuse of double checking, in particular when it is not necessary, had even further negative impact, then the unjustified commitment of excessive resources on the same operation. In fact, this practice had also increased the overall risks on the process, because each operator assigned to the double check was confident that the colleague would have done the check and simply signed the form without any control.

Obviously, the justification of the reasons supporting the modification of the existing procedures, and the reduction of the number of documents and registration forms, was not easy to explain to the operative personnel. In particular, it was stressed that the elimination of a certain number of registration documents was not aimed at a pure simplification of the process, but at the rationalization of the QS, and the purpose was to improve and not reduce its efficiency and effectiveness.

This experience was a good opportunity for me to reflect that the decision of assigning the construction of a QS to an external company, obviously far from the specific reality of the site, should be pondered and evaluated very carefully. There is a lot of added value when the QS is developed by the site staff and is structured and developed in such a way that is dimensioned and calibrated on the context in which it will be applied. In particular, this approach will avoid the risk of building a complex and bulky structure to manage. I believe that the external company, when involved, can have a role of support to the competent functions who, anyway, must remain the first actors in the construction of a real, effective, and robust QS. In the absence of an internal participation, the results may fail to fulfill expectations and oblige the site to commit significant time, human and economic resources on QS modification at a later time.

Examples and Situations That Have Required Mediation and Perseverance Skills by the QA Team

Despite commitment and motivation, the QA team is not always able to achieve what was planned, or in the expected time. However, this may not be its own responsibility. In some cases, when other parties or external CMOs are involved, a lack of organization or misunderstanding with the counterparty can play a negative impact on a timely and effective outcome.

I would like to mention two real examples.

In the first one, the problem concerned the release of the products.

After many years of GMP applied to radiopharmaceuticals, it is obvious that the release of the product must follow a process involving at least three figures in sequence: the production and QC managers and the qualified person. Nevertheless, this was not so clear to some radiopharmaceutical facilities that followed the old habits for product releasing. In the past, the shipment of radiotracers was authorized by a verbal approval of the result of the chromatographic analytical tests, and all data would be recorded on forms and documents in writing only afterwards. This wrong use was based on the justification that the rapidity of the releasing would better cope with the short half-life of the tracer. Obviously, the risk of loosing the traceability of data was immediately evident to QA team.

Changing this habit was difficult, even more than expected. At first glance, there was a generally tensed operational context at the site. It was also evident that there was a lack of coordination and communication between production and QC departments. The alignment of shipping operations, with the verification and approval of the results and their registration were almost inexistent. We raised our concerns about data recording. The discussion between QC operators and QA soon concentrated on who should be blamed for delaying the release of products when "all data had shown that the process went well."

However, we were not discouraged by the low level of the discussion. It took a long time, an intense training of the staff, and the support of the management. In the end, we sorted out the definition of specific procedures, which were widely discussed and agreed upon, and arrived at a perfectly documented release process, working in real time.

The second example concerns the production of the radionuclide needed for the synthesis of the radiotracer. The production was made at a cyclotron, acting as an external supplier, by the bombardment of isotopically enriched water samples; the technical staff at the cyclotron and the facility itself were used to a very light concept of activity tracing, nor used to the timely filling of registration forms.

The enriched water was a raw material subject to analysis, approval, following specific tests, which should have been documented in a rather simple form. The water was under custody of the cyclotron's staff; therefore, we requested that storage and sampling of lots were performed and documented according to our forms. It was a very plain operation, nevertheless obtaining the correct handling and the documentation of the raw material from the radionuclide supplier was very difficult, and company QA had to "fight" a very firm reluctance of the cyclotron staff to cooperate.

In the end, we reached an agreement, which was commensurate with the risk: the cyclotron staff would fill in the forms, but our QA staff would make a periodic check to verify the correct and complete compilation of documents. Of course, a training was essential to reach full operational status.

In the end, through an agreed compromise, the traceability of the operations regarding the entire management of the specific raw material was achieved.

The examples I have indicated so far have a common denominator: training and mediation.

Those who work in QA know very well that any change takes time, and when it depends on the availability of people it can never be planned in detail.

How to Develop an Effective Interface, in the Context of Multinational Companies, in Which the Local QA Manager Works in Close Connection with the Global QA

Typically, global and local QA functions work together on the basis of written procedures that define the mutual responsibilities regarding the management of QS major items: complaints, pharmacovigilance, suppliers qualification, and audits.

A quality policy is defined within the corporate, which is applied to all subsidiaries, and it is the responsibility of the global QA to provide general guidelines concerning specific operating procedures.

The scopes of the global QA are many and impacting.

(a) Standardization of guidelines. This aims to their compulsory application in all the territories in which the company operates. This allows a capillary control, without dispersion of excessive resources in "deciphering" the peculiarity of each single production site, while, however, respecting the specific local legislation.

(b) Simplification of changes. Since processes are managed by the same rules, their change can be managed and harmonized by global QA functions, quickly and in easy way. The analysis of changes that requires alignment among all production sites is developed through the contribution and experience of the entire network.

    Diversification of local activities, which in theory could be an element of attraction due to the different needs of customers in different countries, in my experience can instead constitute an incredible source of complication. The need to satisfy quality requirements, in radiotracers, is so important that assigning complete autonomy to each individual PET production site, can risk compromising pre-established quality standards.

(c) Economic saving. Few procedures, ruling over all processes, well known, and correctly shared by all the subsidiaries mean less efforts in terms of personnel and time to check their correct application.

(d) Training. Strict control and monitoring of training plans is fundamental, better if performed through centralized IT systems with the possibility of carrying out tests and checks remotely.

The standardization of procedures by global QA and its sharing among all subsidiaries have proved to be a good tool for the optimization of common processes, like the surveillance of suppliers and their qualification.

In the quality area, the presence of a limited number of suppliers of critical components, whether they are raw materials or spare parts, allows to have time to qualify alternative suppliers. This is a point that everyone recognizes as a priority but on which little time is generally spent. In my experience, in fact, due to the pressing routine, the requirement of an alternative supplier is often considered only when emergency conditions force the company to move quickly and in a hurry, sometime in a desperate attempt not to block production. Typical examples of emergency are when the supplier decides to stop production or in the event of a shortage of raw materials.

However, there are some aspects, in the relation between global and local functions, that may represent situations that are not easy to solve.

For instance, the time of response by global functions sometimes cannot be in line with the local needs; and exceptional locally proposed actions in response of specific requests from teams or customers, may face a too long time to be discussed and approved at the global level.

This situation can often cause a lot of stress to the local QA, who has to manage in the meantime the pressure of the team that raised the proposal of change. It should be kept in mind that the requests raised require in many cases an important regulatory review and possibly considerable costs for their application.

I would like to mention a few examples of requests submitted by a local commercial department over the years, which surely are frequent and other QA colleagues may also have faced.

(a) Changes in product concentration with respect to that indicated in the MAA. This is to allow, depending on the geographical distance of each customer from the production site, an easier and safer use of the radiotracer by the customer.

(b) Variation of the dimensions of the primary/shielded containers or modifications of the composition of the primary packaging components, like caps, with the scope to make them more easily usable with the instrumentation of the hospital.

(c) Customer requests for product release documentation, such as certificate and injection authorization that are more detailed, in terms of technical information, than company standards.

(d) Request by the customer to include additional information, beyond the one issued globally and already present in the shipping documents (such as references to the order and/or specific codes).

5.3 PET Site Network: Skills Development, Relationship with Management, Performance Monitoring

In Managing a Network of PET CMOs, Did You Have to Develop the QA Skills of Personnel Without Experience? What Difficulties Did You Encounter?

Indeed, in the majority of sites I had to deal with personnel who had no or poor experience in quality management.

After the development of the first PET production site, that was managed directly by the global functions of company, a network of manufacturing sites was our next target. Therefore, commercial contracts were finalized to the production of the tracer with different partners, acting as Contract Manufacturing Organizations (CMOs). Obviously, we represented the main reference for the development of the CMOs' PET manufacturing sites and personnel skills.

Indeed, the experience acquired in building our first PET site has largely facilitate this task.

A multidisciplinary team, constituted by QA, production, and QC, has been at the forefront on the CMO sites since the early stages of the project, for immediately supporting the local team. This allowed the CMO to develop awareness of the QS requirements and ensured the presence of dedicated QA personnel since the beginning of the project.

It is not a condition given for granted. As it has been already mentioned, when a site has to be built from scratch (a common situation), priority is often assigned to solve the engineering aspects, while quality is generally perceived as an element to be included in a second time, when the site has been already structured. Nothing is more wrong!

I was very lucky in the relationship with the teams at the CMO sites: despite not having had a role in the selection of local QA personnel, all the colleagues I met, always young and with little experience, were highly motivated and valid resources to work with.

The lack of a hierarchical constraints with the local team was not a problem in this case, but an opportunity instead; the desire of CMO personnel to acquire the knowledge of QA principles, as well as the responsibility that was assigned to them, acted like a glue on spontaneously collaboration between the parties; perhaps stronger just because born in absence of a hierarchical organization.

In our collaboration with the CMO, we experienced that some difficulties in mutual interactions generally emerged when additional investments, not initially planned, were requested for resources and equipment.

In this case, a far-sighted approach by company QA is needed; remaining firm on extra-project requests and stressing their regulatory implications, without listening to CMO's reasons, can risk creating complex dynamics, and discussions that could have an impact in the subsequent operational phase, in which, even more than in the design phase, a clear, reciprocal, and continuous flow of information is needed.

A practical example could be a discussion that in our experience had occurred regarding the installation of a second independent measuring instrument inside an autoclave. It was an established practice in other countries; however, its need and the resulting economic impact did not always find a prompt agreement by the CMO's QA and technical staff.

Our approach to find a solution, rather than push forward an unconditional request, was based on the provision of data from other sites, which demonstrated the usefulness of this double check. In particular, the double check could have allowed to re-analyze the data in case of problematic production situations and possibly lead to the release of lots that otherwise would be rejected.

Being able to provide the CMO with a technical and tangible reason for the request, like the advantage of possibly "save" lots, was the key to convince them to accept initially unplanned expenses.

Should I have to find differences between being an "internal" QA and a reference QA for the CMOs, I would say that the latter opens to the comparison of different realities, allows a wider access to data, and their analysis when there is a need. It is surely a good opportunity of professional growth.

In fact, PET CMO are often small sites located in various contexts, public or private frameworks, with different policies and regulatory problems. However, they all are required to follow GMP, and sometimes the application of these rules to some organizations is a challenging task.

How Did You Develop the QA Skills to Support the Creation of the QS of the CMO PET Sites?

The training model we adopted was mainly based on interactive and participative activities in the site routine work. Whenever possible, we proposed to QA supervisors and technical staff of the CMO sites to spend a period of about 3 months at our PET site. The activities to carry out during this period are listed below.

(a) Theoretical training on GMP. The duration of this course was defined in consideration of the level of technical knowledge and previous experience of the personnel. The knowledge of GMP requirements is obviously necessary, but above all it is important, through on-the-job training, to see their application in an operational context in which the good practices are routinely applied during production and control activities.

(b) Participation as an observer in the day-by-day QA activities was very useful for the trainee. For example, the possibility to follow how true cases of planned changes, corrective action implementation was managed, and complaint investigation reports issued. This exercise was also the occasion to focus attention on the details of the real cases that were examined and learn the correct recording of the activities performed, indeed a very important aspect.

(c) The participation of QA team in the discussions with the local engineering department during the design phase of the site had many positive influences. A first consequence was that the structure of the procedural system of the future production PET site was beginning to take shape, autonomously and spontaneously, with the contribution of production, QA and QC. In parallel, the CMO team learned how to avoid excessive technicalities in the application of quality requirements and follow only solutions that were commensurate with the actual reality of their site. This interaction helped the local team to identify critical operations, which should be described by procedures, without getting lost in useless documents, descriptions, and details.

(d) Sharing the analysis of real deviations, such as those emerged during the regulatory inspections at other PET sites, obviously taking into account the confidential nature of part of the information, was also a very useful training.

The training period was useful for the CMO not only to learn the basic aspects of the QS, but also for us to build a relationship based on mutual trust.

Once the training was completed, the CMO QA supervisor and technical team maintained a constant, almost daily, communication with our QA team, thus allowing a natural and friendly way to remain updated on the PET site daily operational. Years later, the people we have trained, even after moving to other pharmaceutical companies or having reached top responsibility, have continued with this practice. This is a proof that the "method" worked for their professional progress, even beyond our initial objective.

What Process Do You Think Should Be Adopted to Monitor the Performance of a New PET Production Site?

During the early period of production, it would be appropriate to schedule frequent discussions, not less than once a week.

When more practical experience is acquired at the CMO site, the frequency can be lowered to a process based on written communications and periodical reports.

Obviously, deviations or changes should be promptly communicated by the CMO, along with any reports produced and subsequent updates to maintain full track of the issue.

In fact, the CMO site produces on behalf of the MAA owner, who is the final responsible for the quality of the product under any circumstance. In this perspective, the definition of a quality agreement that establishes the reciprocal responsibilities of parties is of fundamental importance.

What Difficulties Did You Encounter in Dealing with the CMO Top Management?

The most evident problem with the CMO management is the initial underestimation of the difficulties that the site will inevitably encounter in implementing its own QS.

The typical statement is "there are no problems, we have the expert people." Generally, after the start of project activities and after facing the real problems of qualification/validation activities, their position decisively changes.

Following this early phase, as I previously mentioned, the next possibility of stressing situations is in relation to the level of modification needed vis-a-vis the initial project, or to the planned time schedule. In other words, anything that involves an increase in operational time and resources, and therefore costs, may lead to a somehow conflicting discussion.

The inspection by the regulatory bodies to authorize production or by corporate for the approval of the CMO site as a supplier are all elements that may stimulate a collaborative reaction by the CMO on investments. When the site enters routine production, things get more complicated. In fact, changes that go beyond the authorized situation can be perceived not urgent and even unnecessary by CMO top management, an hurdle to the normal production activity. Here, the role of company QA is crucial, in trying to carry out an objective analysis, by means of a technical and assertive approach with the CMO, of the advantages and opportunities linked to the change.

5.4   Organization of the QA System in a PET Pharmaceutical Facility

The Procedural System: What and How Many Procedures Should Be Developed, and What Are the Mistakes to Avoid?

In many PET sites, it is possible since the site design phase to identify and define many of the process operating procedures, which will be part of an integrated QS.

The drafting of the procedures is conducted in cooperation between supervisors and operators of the various departments, and it is supported and supervised by the local QA team who will provide correct regulatory references, and homogeneous structure, of paper or electronic forms.

During the project phase, the procedures that are directly linked to the areas under development or construction, and those ruling over changes to the QS should be prioritized: personnel and material flows (from warehouse, to production, quality control, and shipping), environmental controls, use of equipment, deviation, change control, and validation activity.

The supervisors and the operators should identify, together with the engineers, the sequential steps of the processes and the necessary flow of information. Starting from this, they could prepare the essential working documents and registration forms, which may include a concise descriptive text for each activity and points of data registration. Linear schemes and process flowcharts would be of utmost practical use at this stage.

These schemes can replace too detailed descriptions that the operator will hardly read, and remain more tempted by memory and verbal indications of "expert" colleagues.

I will never stop to stress that training, on the correct use of registration forms to anyone assigned to production, is of vital importance for a correct passage of information (transparency of processes) and prompt problem analysis.

Errors can unintentionally occur and lead to an unbalancing in the QS, both in terms of number of documents and complexity of the structure. Below are listed the most important points to consider while setting up a procedural system.

(a) One of the most common mistakes is creating blocks of activities that are extremely detailed but with little or even no cross-connection.

Let's take, for example, the tracking of a raw material that is moved across many steps of the production process. A strict cross-reference among the documents ruling over the process is very important to guarantee the raw material correct handling, its identification and localization inside the process, at any step and moment.

Likewise, if a raw material is listed in the storage records of the warehouse, there must be the reference to its analytical sheet, including the minimum tests necessary for its approval, the definition of the minimum stock volumes and its use in the production process (Master Formula). If the analytical tests are performed internally, there must also be a reminder to the QC methods and equipment used for the indicated analysis. If the tests are instead performed by an external service, the company providing the service must be qualified, and a contract set up, which covers the responsibilities of both parties and the procedures for the reporting of test results.

If such a documentation is not present and correctly cross-referred, we can probably face a lack of information and, in the end, a situation in which not all aspects of the process are covered by the QS, that would not result complete and robust enough.

(b) The procedure should be prescriptive and prone to keep errors to a minimum; they must never give rise to interpretations.

The PET manufacturing world is characterized by a high degree of working autonomy in the staff. The number of personnel on staff is generally small and, as a consequence, extremely specialized. This condition and the psychological pressure posed by the sequential manufacturing of lots in a short period of time (few hours), limit the possibility that the staff on duty has to interact with colleagues and supervisors, and discuss on doubts or abnormal situations that may occur during their night activity. Following this peculiar working condition and after having acquired some experience, a common staff attitude develops: the operators do not refer with the due attention to procedures, but tend to carry out activities relying on their own memory. It would be good practice, instead, to keep procedures and worksheets at hand during operations, as they are the primary tool for reducing human errors.

This is demonstrated by the fact, confirmed by experience, that the trend of this type of error tends to increase when the production site reaches a certain level of maturity. On the other hand, it is more natural for the newly recruited staff to look at written procedures during the start-up phase, when they have to develop manual skills and pay more attention to reduce the risk of errors.

(c) Procedures should be modified/updated by the same personnel who elaborated the drafting.

The update of the documentation system (procedures and forms) is outside the specific competence of QA. A mistaken belief is that the QA team should be himself responsible for procedures updating; in my opinion, it is completely wrong. The QA team needs to have the documentation system under control, including the check of validity as defined in each procedure, but it is the responsibility of each specific department to update their own documentation in line with the changes that are made to the processes, due to regulatory or any other valid reason.

The continuous training by QA and its presence in the different departments will provide help on the correct use of updated documentation and play a surveillance role on changes versus supporting data.

It is normal, especially when the site gains operational experience, that the initial document templates may become obsolete or not fully compliant with operations. In that case, it is necessary to modify or realign them by means of a controlled change process, which includes a risk assessment before the final authorization for implementation could be finalized.

With the start of PET site routine operations, minimal changes, sometime made by operators themselves in autonomy, are not considered worthy of a risk assessment with the supervisor before being applied. If the QA does not timely intercept these situations of misalignment, because the communication system is not sufficiently effective, the major risk is that changes missing a controlled change process might become a normal practice, and the operators will start to follow informal and random directives rather than up-to-date approved procedures. It would be a very dangerous condition with heavy consequences on product safety and site operation.

Change Management

What Is the Role of QA in Implementing Operational Changes? What Are the Difficulties Encountered in Managing Changes on CMO Sites?

In my opinion, any change decided by the company, either organizational or regulatory, should be jointly discussed in meetings involving the different functions at the site: this is a good strategy to evaluate their impact on the global process and on the QS in general.

The process of change control should be coordinated by the QA with the aim to guarantee that all modification impacting on the QS are correctly considered under any aspect:

(a) Risks involved in the change
(b) Possible extension, in terms of technical, logistical, or documental aspects
(c) Regulatory compliance and needs of authorization

This coordination should also include the control over the implementation time, as well as the verification of effectiveness and validation when necessary.

Following our experience in the PET area, operational changes and implementation plan are decided with tight timelines, predominantly due to economic strategy and/or technological updating pressures. Common situations at all PET sites vary from the replacement of critical instrumentation or logistical system, to the adoption of modifications aimed at optimizing processes or increasing productivity.

The construction of new production lines, the partial upgrading of the premises, and the installation of new equipment can be a much more challenging decision.

However, the common denominator for any kind of modification is "to act quickly and finish soon," without interrupting production activities too long.

The role of QA, regulatory, technical and radiation protection functions is the evaluation of any possible impact of changes on the production process, as well as on safety and authorizations (in terms of radiological license and MAA).

Changes to the MAA are highly impacting, especially in terms of time for approval; therefore, the analysis of a change, by all the functions mentioned before, is a necessary step for the company to have a clear and complete picture of timeline, risks, and costs linked to the project.

A more limited analysis, for instance, left only to the evaluation of the functions proposing the change or an external service, may risk to end up into unexpected regulatory complications instead of saving time, and consequently involve longer times of execution and health authority change approval (if any).

The coordination activity of QA consists of collecting the contributions from all functions to perform the analysis of modifications, including considerations on site structure and process, and any other pertinent QS implications. For example, a technological or structural modification of a PET site is often linked to radiological aspects, beyond those related to product distribution and logistics.

The QA team should also closely supervise the planning of the change and the preparation of all the supporting documents, validation activities, and standard operational procedures (SOP) to be used after the implementation of the change.

Sometimes QA and RA can enter a conflict with the company's expansion projects. This can happen, for example, in consequence of the company request to increase PET site manufacturing capacity, by increasing the produced amount of radioisotope used for the radiotracer synthesis, so that larger batches can be produced.

Apart from possible limitations imposed by the radiological license of the site, this is not a minor change, because process and procedures only apparently remain untouched.

Exceeding the production limits approved for the site, which are reported in the relative MAA, implies the revalidation of the entire process. For instance, the stability of a radiotracer can be affected by the increase of total radioactivity, due to enhanced radiolysis problems. Likewise, the sterilization process, whether performed by autoclaving or filtration, could require an ad hoc validation if product concentration, number of vials, total volume to be filtered are different from the approved process.

Should process revalidation and MAA modification be the way, QA and RA will play a fundamental role to prepare the application, in terms of coordination, analysis of the technical changes and time planning, and ensure adequate follow-up during regulatory approval and site implementation.

Global QA can act as a reference, with a supervisory role, on the management of changes to be implemented at CMO network sites; especially during the initial training phase, the local CMO QA staff must be made aware that changes have unavoidable GMP and regulatory implications.

The changes concerning MAA sections are coordinated by the company that owns the authorization, and the main difficulty consists in the alignment of the implementation of changes throughout all CMO sites. It may be a very cumbersome task. For example, changes on product labels and their harmonization across the different countries required a relevant effort in terms of implementation timeline, from both local CMOs and company sides (review and approval), and homogeneity of format and content due to different languages and printing systems.

An Example of a Change That Is Extremely Difficult to Deal with

The centralization of purchasing and supply of raw materials for production we performed in a PET manufacturing network showed advantages and disadvantages. Here, the "network" is the pool of PET sites that receive the supply of production materials from one "reference site."

This decision was driven by the need to optimize stocks and use of external services for chemical and microbiological analyzes, and their relative costs.

Our experience as reference site for the centralization of supply included both raw materials and primary packaging. We had to consider and solve some critical points. For example, the direct delivery of raw materials from suppliers to the other PET sites in the network was not possible because the analytical tests, performed by the supplier, were not accepted by all countries.

A possible risk associated with a centralization strategy is that one critical issue, such as in the case of material or test analysis problems, would impact on the entire network. Therefore, centralization requires an in-depth analysis of the QS to be developed for both the reference and the satellite PET sites, to construct a robust supply network. Some key points we had to address are summarized hereunder.

1. Consumption forecasts. This aspect includes the implementation of a robust and strictly programmed communication system within the network. It must be structured in a way to constantly monitor consumption and needs of raw materials at the various sites. It is easy to say but, quite complicated to make it real. The production of raw materials (primary glass containers, reagents, and purification columns) was often planned by original suppliers based on advanced requests, to be updated at least yearly. As reference site, we were often in the position of having to balance the pressing needs of the supplier to receive information with the due advance, and the delay of the network satellite sites in presenting their planning.

2. Documentation language. The operating procedures and the related forms must be modified so that they can be easily understood by all the PET sites in the network. Thus, English was used not only in the final certification documents, but also in the labels accompanying the raw materials. This aspect should be carefully considered in the initial selection of suppliers, because not all suppliers being contacted were flexible enough in accepting our language requirement.

3. Packaging and shipping. A multifunctional collaboration, involving technical and logistical functions at our reference site, was necessary to look for suitable packaging materials, provide specific control systems, and to monitor critical shipping conditions. For example, certified and calibrated data logger were used for controlled temperature dispatch of products and for giving evidence of temperature maintenance during the transport to the destination satellite sites. We had also to organize additional space and dedicated storage locations to stock the material for all the PET sites, waiting for shipment and delivery.

4. Supplier. The selection of suppliers, in most of the cases, was done in collaboration with global QA, because the specifications of raw materials were defined in the MAA. Our main commitment consisted in the continuous surveillance of supplier's performance, by means of intense auditing activity and punctual control of the materials provided by each of them. As a reference site, this situation involved a big change in our QA activities: the audit programs were extended, and the number of personnel dedicated to audits and reporting activities increased as well. The preparation of the audit check list included a broader view of the points to be verified, which included the collection of observations from all PET

network, thus maintaining a most possible shared evaluation of suppliers' performance.

5. Raw material specifications. What is accepted in one country may not be aligned with the needs of other territories. The distribution of products in various countries required the harmonization of all analytical tests and regulations applicable in each individual receiving country, and therefore reported in the certificate of analysis for release.

   The inclusion of the various specifications in a single release certificate, required a comparison between the QA/RA/QC functions of all the countries.

   Eventually, the final list of analytical tests, as well as their validation/revalidation, was incorporated in the specific agreement made with the supplier of the analytical service.

6. Effective communication system. A robust communication system, throughout the entire PET network, was set up to develop an effective information channel about the consumption of raw materials and the resulting orders to place with suppliers. Various activities were included into this communication system:

   (a) Collection and tracking of issues directly related to any raw material
   (b) Management of logistic aspects connected to specific regulations on material transportation
   (c) Management of raw material delivery times, in line with production needs
   (d) Tracking of raw material data, to guarantee the correct retrieving of information, in case of any adverse events investigation

       All the above-mentioned aspects required specific procedures, which were agreed and shared with the various countries. Of course, the preparation of these procedures included the contribution from multidisciplinary expertises (logistics, QC, RA, production, QA) collecting the needs of all network PET sites.

7. Staff. The new commitments resulting from the centralization activities within the entire network required a complete retraining of operational personnel, and an adjustment of the internal work organization at the reference site.

   This change was very complex and followed a gradual implementation, due to the multiple actions we had to apply to the entire system.

Have You Ever Experienced Changes That Were Not Implemented as per the Original Plan?

As an example, I would like to mention a change of plans that turned out to be much more complicated than expected and ended up in an alternative decision. The need for a change was triggered by the notification of problems at some manufacturing sites, concerning the primary product packaging. The problem occurred during the insertion of the glass vial, already loaded with the radioactive product, inside the shielded container, and was so severe to jeopardize product delivery. After an in-depth data analysis, a candidate cause was postulated, even if it was not possible to absolutely establish its clear connection with the problem reported.

The discussions with the supplier considered a possible radical solution of the problem, by introducing changes in the primary container. However, this solution was very complex from the regulatory point of view and there was no proven evidence that, once adopted, it could have been resolutive.

Thus, we continued to analyze in more details any single segment of every step of the entire production process, trying to isolate all factual contributions that could have impacted on the occurrence of the negative event, even those coming from the equipment and the method applied.

By acting on these segments, we added specific precautions to avoid mechanical stress during vial handling in production, correct storage, and more suitable packaging assembling, and finally we reached the scope of almost completely eliminate the problem occurrence.

This is an example of a major change, initially considered unavoidable, that was not eventually applied. The experience taught us that a change that could immediately appear as decisive, just because radical, on a further analysis, turned out to be not only potentially unsatisfactory but also more expensive and complex than carefully calibrated minor modifications.

Supplier Management

What Is the Role of QA in Choosing Service/Equipment Suppliers and Their Qualification?

The qualification of service or equipment suppliers should be a QA competence, conducted in cooperation with technical and engineering departments. Unfortunately, the QA is often involved only at a later time, when the supplier has been already selected.

Limiting the number of qualified suppliers, with well-known history and level of performance, to a minimum would be a company added value and could speed up and save time and costs of projects, e.g., of other PET sites, in which the same suppliers can be used.

The equipment used in PET manufacturing requires particular care and a forward-looking strategy. In fact, some critical equipment, like that used for radiotracer synthesis or dispensing, is accompanied by specific consumables, in the form of cassette, special reagents, and kits. In this case, the qualification should be extended to subcontractors (if any) supplying such consumables, and it may happen that they have to make additional effort or even small changes to their production process in order to align with GMP requirements.

For the reasons described above, it would be advisable to keep the list of approved suppliers always updated in terms of qualification, requalification, and surveillance.

Furthermore, it would also be helpful considering the effort otherwise needed in this process when company or CMOs needs to look at equipment replacement and technological development.

Management of Deviations

What Is the Role of QA in the Management of Deviations at a PET Site and How Do Other Functions Contribute?

An anomaly or a "deviation," in general, is an activity that deviates from an established scheme or a result that does not meet the expected specifications.

The deviations potentially affecting the quality and safety of the product are regarded at a PET site as events to be solved as soon as possible, if not immediately, due to the short production time frame.

The management of deviations, likewise all aspects of a QS, depends upon the training of personnel and includes the immediate involvement of supervisors and QA.

The QA team once again acts as the coordinator and manages a process including the following activities:

- Deviation recording
- Meeting organization, focused on deviation analysis and involving the competent personnel
- Coordination of the investigation
- Advice to the supervisor on immediate actions, such as rejection or recall of a defective batch
- Coordination in the definition of medium and long-term plan of corrective and preventive actions
- Verification of corrective/preventive actions effectiveness

All these actions are the same in any QS, but the peculiarity of a PET site is the necessity of their implementation in a very short time, in line with the speed of the manufacturing process.

A particular case of deviation is the "programmed deviation" from an established process/activity. For example, a "temporary change" or "short term change," with a limited defined time of application, can be applied to a process development. The underlying assumption for the adoption of a programmed deviation is that there must be, in any case, no negative impact on the safety and efficacy of the product. In principle, it can be applied in any area of the QS.

The QA team opens a change report to keep its application under strict control, with particular attention to entire process traceability, including risk analysis, change approval, and any connected corrective actions. In this situation, the close cooperation and the relationship of trust between QA and the involved departments are essential conditions for the correct implementation and success of the planned target of the programmed deviation.

Failed product release is considered as a deviation, and indeed they are the most difficult to manage at a PET site.

The immediate complication is the direct inconvenience to patients, who have a treatment programmed on the specific day of the failing batch. On the other side, there is the prompt reaction by the customer, who will question service failure and, should a failure have occurred again over time, site reliability.

Obviously, this particular frame of events brings the problem to the highest level of priority in the company that will require the site to find an immediate solution. The site has to fix the situation in a very short time, but after having solved the root cause of the failure and been able to restart production in full safety.

Therefore, the QA team on site must open an investigation as soon as possible, together with the technical functions, and in case of a CMO, must alert company QA on the current occurrence. Whatever the case, the top management will expect prompt evidence and solutions.

A possible reason for a failing batch is the presence of an Out of Specification (OOS). An OOS can be related to chemical or microbiological problems occurring in the process, so that the final product is outside the approved characteristics for a safe and effective use. Another possible failure depends on a defective performance of the automated equipment used in the processes.

The evidence of a failure in the production equipment or of a clear human error, even if undesirable, represents an immediately detectable problem and often simplifies root cause identification and deviation analysis.

The investigation of an OOS when equipment reports are compliant, and process apparently performed normally can be very complex. A PET process covers a long way from production of a raw material (the radionuclide), to synthesis, purification, packaging, quality control, and shipment. Sometime the root cause is so elusive that the advice of specialized technicians or product experts can be necessary. Unfortunately, the intervention of specialists may not be immediately available, or the corrective action may require a requalification of the equipment: too much time taken off from production, distribution, and patient treatment, the three key phases of the PET process, whose speed and correct sequence are crucial.

In our experience at PET sites, human error in the set up of the synthesis or the dispensing module, such as mistaken connections or loading of reagents, is one of the most frequent causes of production failures. They can rarely be detected by the software during module operation or identified in the printout of the module working parameters afterwards.

Human errors are hard to investigate and resolve. Furthermore, employees are often reluctant to admit or openly recognize that they made a mistaken operation; therefore, they must be convinced of employer's positive attitude, without the fear of risking their job, and its ability to deal with the negative event and resolve the problem in a constructive and inclusive way.

Unfortunately, the pressure on the production area, brought into play by all the other company sectors, like commercial and distribution, does not often contribute to create an atmosphere of collaboration, and facilitate the admission of a personal responsibility, in particular when the deviation, such as the rejection of a batch, involves an economic loss for the company.

What can QA do to mitigate this issue?

Certainly, QA should insist on the need to have a robust process during training, making it clear that error analysis can even imply a critical re-evaluation of the entire system and remove hurdles that may induce mistaken operation.

In general, the analysis of a human error is carried out by focusing on the most critical points of site organization and PET production, such as:

- Too much compressed working times during production
- Understaffed team versus operational needs
- Potential weakness and need for modification of the procedure(s) involved in the negative event occurred
- A need for further training of the entire team, and not just the sole individual operator(s) involved, to develop a more robust working method, and avoid the reoccurrence of the anomalous event.

It is important to consider that there is a strict link between pharmaceutical operations and radiation protection in a PET production process; a single error can lead to an important risk for the safety of employees and, in the end, heavy consequences for the company.

Internal and External Audits (PET Network CMOs)

What Is the Added Value of a Periodic CMOs Audit, Both Internal and External, as Compared to the Daily Routine Surveillance?

Internal Audits

Self-inspections are generally organized in all departments in rotation to monitor the effective functioning of the QS during the calendar year.

In my opinion, the definition of the audit relative checklist in an ongoing and operational QS, should be based on specific aspects that can provide evidence of the effective functioning of the system.

For example, I would take into consideration some specific aspects over a certain period of time, such as those described below.

1. The methodology used for managing deviations and the role/involvement of each function in activities like the start of the investigation process, the tracking of results and the ability to keep pace with the planned time frame.
2. Deficiencies of communication among areas in handling change controls and complaints.

Internal audit should be a tool that allows to verify and constructively investigate what may become "jammed" in the QS, compromising its regular functioning and the ability of tracking processes.

The audit must be conducted in such a way as to guarantee a useful collection of objective evidences, which can be a starting point for improving the functioning of processes in the various departments.

External Audits (PET Network)

As already mentioned, PET CMOs have always been supported by a company multidisciplinary team (production/control/QA), during both site construction and production start-up.

Once the production licence has been granted by the competent authorities, the CMO becomes independent and autonomous in running the process and operates with its own QS.

In general, the relationship between the company and the CMO is regulated by QA agreements that establish the responsibilities of the parties and their connection/communication points by means of reports and inspection plans.

We approached inspections of CMOs as a contact point, aiming to discuss the problems that had occurred and the corrective/preventive actions that required a common evaluation.

The relationship with the CMOs was always based on absolute trust and collaboration during inspection activities, and I have always supported the idea that audits were performed by the same multidisciplinary team that had supported the initial start-up phases. There are good reasons to have this approach.

- The technological component of a PET site is extended and important, and an in-depth knowledge of the site will facilitate prompt interaction during the audit. The multidisciplinary team of both parties (company and CMO) can support an immediate analysis of any problem identified and can more quickly provide an analysis of the actions that have to be taken.
- The presence of experts in the company team of auditors promotes a more efficient interaction between peers with the CMO team, in terms of competence, participation, and motivation. It will also allow a more integrated vision of purely technical and quality management aspects as a whole.

An important indicator of a CMOs QS performance, that should be observed during an audit, is certainly the analysis of complaints management. In fact, the verification of the investigation mechanisms and the communication flows adopted by the CMO will provide full evidence of the correct interconnection established between CMO departments and company equivalent functions.

The effective tracking of all operations unavoidably emerges during the analysis of complaints, and this process, if well managed, represents an important moment of confrontation between the parties, and a starting point for improvement as well.

Complaints, Ideas for Improvement

What Is a Complaint? Should All Reports Be Logged as Complaints?

A complaint is the external notification of a potential defect of the product or of the process, that is received from a customer. No process is free from potential problems, despite having been standardized even in the smallest detail.

I am convinced that all customer notifications should be recorded, evaluated, and eventually identified as a complaint.

Notifications sent by customers are often numerous; in general, not all of them are necessarily to be considered immediately as complaints. Sometimes, they only are requests of information, clarifications or even support, for example, about product preparation. Conversely, the high frequency of a specific request, initially not considered as a complaint, can raise concern and be investigated as such.

The person who receives the notification, and even better who analyzes it, should carefully evaluate its nature and make a correct classification.

In our PET experience, the complaint may address any step of a complex operational process (order entry, production, delivery of product) and the employee who first receives the notification may not have the competence to do an objective analysis of the problem reported.

Complaint evaluation can only be done by a competent function through a technical analysis. For this reason, the definition of a robust communication process among the areas involved is of fundamental importance for the correct management of complaints. Key points in this process are listed below:

– Define specific procedures for passing information across the departments
– Receive active collaboration by functions and operators in the investigation
– Ensure immediate retrieval of information and data that are needed to evaluate the problem notified

Often, there is a wrong attitude to keep relevant information on the complaint only limited to the individual function, due to the belief that outside there is no competence to judge a specific technical aspect. Instead, only the comparison between different departments can lead to a broader and more in-depth picture of the possible causes of the notified occurrence, which may not be identified by a sole individual sector.

In very small organization, such as a PET site, it is often possible that complaints can be verbally notified by the customer, to a driver of the delivery service or a production operator. Therefore, it is important that all involved personnel be trained on the correct channeling of complaint information to the competent function, the QA team, which will start processing by complaint registration and investigation coordination.

The initial investigation step is the identification of the team that will be involved in the analysis of the complaint. In many cases, the team includes a distribution representative and experts on transportation problems. Logistical problems, such as road traffic, efficiency of vehicles, adverse weather conditions, can heavily delay the delivery times which, considering the short half-life of the product, are crucial for the customer to keep the actual patient treatment schedule.

The investigation team can be extended when suppliers used by different CMOs are involved in the same kind of complaint. In this case, the team should include also representative of suppliers of raw materials and/or transport services.

Potential problems may not raise only internally, but also externally, from one of the many suppliers that are involved in PET production. For example, small changes made by a supplier in its internal production process, considered riskless, can generate even critical problems for PET production, because of the complex chemical and technical nature of the process.

A potential defect notified in a complaint could, in some cases, constitute a faint alarm signal of an aspect which, only following in-depth analysis, may lead to the detection of a serious threat on production, or a sign of a negative "trend," which is gradually emerging from several correlated notifications.

Therefore, the complaint, even if obviously not desirable, should not be regarded only as a negative thing to be quickly settled.

In a complex context, such as PET productions, based on a fast process, once the cause of the complaint is verified and defined, the corrective actions (often including mechanical and software modifications) can require long times for their implementation and, sometimes, authorization by the competent authority. Thus, having found the cause of a complaint, the site should move quickly to perform any necessary analysis of due changes and define the timescale for their completion.

Complaints on logistical problems mentioned above could be an indicator about the effective ability of the production site to address customer expectations, because of its production capacity, or site geographical location.

In this case, unless long-term corrective actions are needed, such as structural changes, the satisfactory response to the complaint can reach the customer in only few days or even be immediate.

Therefore, to reduce problems of logistics, it would be strongly advisable to devise a preliminary alignment between the commercial strategy (for example, number and location of customers in a specific territory) and the level of technological potential necessary to satisfy the strategic targets, and guarantee product and delivery service good quality.

Complaints on product problems, for example, concerning active concentration, dispensed volume, primary and secondary containers, require much more time for analysis, because, if the complaint is confirmed as such, it could be the starting point of medium- or long-term changes.

In PET manufacturing, likewise in any pharmaceutical industry, the worst outcome of a complaint could even lead to a product recall. It is a rare case, which anyway emphasizes the importance of a correct evaluation and management of any notification received by customers, and any problem that can be experienced by customers while using the product.

Finally, the author of the complaint always assumes that a professional expert will reply in a short time, and it is a wise decision that the company always assigns high priority to complaint management. However, the time for closing the complaint, obviously, can vary according to its nature.

What Is the CMO Involvement in Handling a Complaint? What Issues Might Arise?

Inside a PET network, complaint management is necessarily a shared activity between the company, that is the owner of the MAA, and the CMO. This process must take place having first defined the responsibilities of developing the investigation and providing a response to the customer. These responsibilities of the parties are set out in a contract, usually consisting in a section of the QA Agreement.

The investigation effort on the CMO side must cover all production and control data and be extended, both internally and externally, so that immediate and detailed reports are provided to the company, along with any information useful to answer any request by authorities, should this be the case.

An operational method for complaint management should be agreed and should include the initial communication phases between the parties, the investigation, the timing of reporting, and the definition of the response to the customer.

The qualified person has the authority and the responsibility to decide on the severity of the problem and immediately contact the relevant health authorities in due cases. This eventuality requires a very close communication, between the company and the CMO, for the managing of any consequent regulatory and distribution aspects.

Of course, everything is complicated by the very short time scale of the process: only few hours elapse from production to the almost immediately administration of the product after delivery.

Complaint management requires the training of CMO team; however, customers also should receive the correct information about the complaint process. For instance, a negative example in the managing of a complaint was connected to a case in which the notification was provided by the customer directly to the transporter in charge for delivery of the product, rather than to the manufacturing company.

This wrong alternative way of notification, if unstopped, could generate an uncontrolled, unclear, and risky parallel management of the complaint, ranging from the failure in capturing a complaint to the lack of the investigation of a potential defect.

The company and CMO QA teams play an important role in the training also of external personnel in charge of distribution, commercial contacts, and supply of external services. This training must include clear information to the customer, relating not only to products but also to services. Indications must be given of the correct communication channels to be followed by the customers; in particular by those that provide information to the manufacturing site about any product defect, and service deficiency.

Key Performance Indicators (KPIs), How to Monitor the QS of PET Sites

What Is the Meaning and the Added Value in Defining KPIs in a QS of a PET Site?

The key performance indicators (KPIs) to be taken into account to monitor the robustness of a QS should provide information about the effective performance of the mechanisms that globally make the QS a harmonic and functional structure for the organization.

KPIs should be valid for the QS of production sites that are managed both directly by the company and by CMOs.

The ability to meet the planned commercial strategy (number of batches versus customers to be served, amounts of delivered radiotracer) is the direct consequence of a good performance of the organization as a whole and, therefore, of its QS. A good QS performance means qualified suppliers, regularly maintained equipment, robust manufacturing and control mechanisms, and efficient customer services.

An external observer, based on a mere commercial consideration, could errone-ously assume that the production site regularly supplies and delivers all product batches to the customer, according to the production plan and in good time for patient treatment, without any particular interventions. Instead, this strictly depends on the daily work of the QA staff: making all the mechanisms defined in the system work in an integrated manner.

This is achieved by QA by defining, together with the technical departments, KPIs that efficiently measure the real functioning of the processes.

The performance indicators can vary based on the critical points that need to be kept under observation.

Of course, for each KPI, an acceptable risk level is defined as a target. In our experience, the most common KPIs of a PET site are defined in the following areas: deviations, complaints, changes, operating results, supply management, and reten-tion of people.

1. Deviations. The percentage of deviations that have been resolved in an estab-lished period of time can be defined as a KPI. Deviations are generally classified by category, given the various areas in which these can be detected. Obviously, with the progressing of the activities at the PET site and their evolution in terms of production capacity, this classification becomes increasingly broader and more detailed, even if the macro-areas remain the same: production, QC, pack-aging, and logistics.

    Another example of KPI based on deviations can be the number of critical deviations detected during internal inspections or audits by external bodies, such as the regulatory authorities.

2. Complaints. The performance indicators in this area can be defined on the basis of the problems that can arise during the management process, rather than the total number of complaints received:

(a) Number of complaints confirmed as such
(b) Investigation timing, which must fall within a specific time frame, to guarantee the efficacy of the process
(c) Customer response time (sometime, in case of PET network, depending on the effective interface between the company and CMOs).

3. Changes. A KPI can be defined for changes in a critical area, for instance, the implementation timeline. In the specific case of a structural change (e.g., extensions of production areas, new equipment), the timing can only be measured exclusively on company activities, like:

(a) Preparation of project documentation and risk analysis
(b) Execution of the project and qualification activities
   Instead, the time needed for receiving permits or clearance by authorities, which may also include an inspection step, is only partially predictable or amenable to a standard.

4. Operating results. Examples of indicators useful to measure production capacity and maintenance of operational commitment, that is at the end the production robustness, can be the following:

(a) Total number of batches produced
(b) Percentage of rejected batches with related investigations concluded
(c) Percentage of the achievement of minimum guaranteed production capacity.
   Clearly, a negative trend in production robustness can be a sign of weakness in the manufacturing process management, which could require prompt adjustment.

5. Supply management. KPIs can be defined providing, at a single site or within a network of PET sites, the measurement of the ability to guarantee the continuity of both production and distribution.
   In this area, the following KPIs could be used:

(a) Percentage of mistaken plans of the request of raw materials to the supplier. This aspect is particularly critical in a PET network, in which a centralized purchasing strategy is in place. Errors in this activity can lead to possible supply shortages at various sites in the network.
(b) Percentage of suppliers not managed with contracts. Individual/spot orders, rather than contracts, may not cover all the critical aspects of the supply, first of all the delivery time that is fundamental for the regular production continuity.
(c) Percentage of materials and services that do not correspond to those ordered. The impact of this defective service is particularly impacting on PET sites, in which there are not always qualified alternative suppliers able to cover the lack of materials in the short term.

6. Retention of people within the workforce. The ability of retaining people within the PET site organization requires a lot of strategy, professionalism, and sensitivity. The personnel employed at these kind of manufacturing sites have a specific

and peculiar professional profile, often developed following highly specialized training courses, and a considerable financial effort by the company.

At the same time, the work is often intense, repetitive, and performed during night hours. The timing for production, analysis, and product release is very short, and when production failures occur they must be faced in the immediate, trying to find prompt solutions.

It is the duty of site managers to balance the stressing routine with the fulfillment of expectations that emerge in people in whom the company has made substantial investments, in terms of professional development.

The most effective way to achieve this goal is to build, within the site, an environment of trust and mutual exchange, from both a human and professional point of view. Inclusion and promotion of continuous active collaborations can be the tools useful to create this condition, and guarantee the functioning of the QS that governs all site activities.

Examples of critical points to be monitored in this process can be summarized as follows.

(a) Motivation, reinforced by alternating routine and project activities. A KPI could be defined by measuring the number of projects and trainings actually activated/executed as compared to those planned, on an annual basis (example of a minimum target for this KPI: at least 50% of the planned projects/trainings).

(b) Scheduled shifts, which allows for an acceptable quality of life. The KPI to be monitored could be the percentage of overtime hours outside the scheduled shift (example of maximum target for this KPI: no more than 10% of the total working hours planned each month).

(c) Availability on emergencies (to be applied to the whole workforce). The KPI could monitor the distribution of emergencies over the entire workforce (example of target for this KPI: gap among employees not exceeding 10%).

If the mentioned KPIs result outside the defined target, it is possible, as a consequence, that the number of people leaving the company could increase, putting at risk production continuity, system robustness, and commercial strategies.

As similar problems can involve different PET CMO network, some specific KPIs could be shared and monitored, with the aim of guaranteeing homogeneity of development and improvement.

KPIs and their trend must obviously be periodically analyzed, within both the company and the CMO network.

The number and complexity of KPIs generally develop in a progression, especially for newly activated production sites, with limited availability of data or information, that are able to support a useful analysis to evaluate performance.

Conversely, in the case of sites with long experience, it would be useful to focus KPIs on challenging targets, rather than flattening on already consolidated and easy achievable objectives. This attitude will push the site to aim to a continuous development strategy.

5.5 Movie or Reality?

Can It Be Sustainable Keeping a QS That Is Based on a Virtual Structure?

In the effort to give a positive impression during the inspection, a manufacturing site will obviously commit to prepare the situation in the best possible way. Instead, this effort may risk of giving an artificial impression of the QS, purposely built to convince the auditors.

Starting from this thinking, the internal saying that the QS could be a "movie broadcast just for us" emerged as a joke during an audit at a PET site.

With the term "movie" we intended something designed to perfection but functioning only temporarily. A system thus conceived would be antithetic to a QS, a "house of cards" that could fall down disastrously under a careful analysis, raising an embarrassing and depressing situation.

As I mentioned before, in our experience, the audits of PET sites in the network were typically performed by a team in which QA was only one of the member. In fact, technical experts, such as production and control representatives, were also active part of the audit team evaluating the goodness and a truthfulness of the QS inspected.

A QS is a complex structure constituted of well-defined, traced, and documented interconnections among processes, which globally are the gears of the system and can be easily verified during the audit.

It is precisely through the experience, acquired in similar situations and with the same emotional loads, that auditors are able to look critically at aspects that might seem obvious.

We can definitely say, from our experience, that it is quite impossible for an entire system to be kept functional only virtually, intermittently or for limited periods of time. In case someone thinks this is possible, it would need more resources than those necessary to apply it in a normally, daily functioning system.

It should be said that audits, for reasons related to resource optimization and the ever-present need to reduce costs, are extremely limited assessment moments, both in terms of time and materials, environments, and documentation that can be reviewed.

In this short period of time, however, an expert auditor is able to evaluate whether the QS under observation is actually functional and robust or is the so-called movie, set up for the specific audit situation. This can be done by inserting audit specific observation points that will be described below in details.

The virtuality of a QS could already emerge in the context of the presentation meeting, in which normally, in addition to the technical/operational figures, it is important that the company top management is present. Its absence, in fact, would be indicative of a little interest by the top management on the typical mechanisms of a QS, such as the audit.

The presence of the management, and its opening presentation, allows the auditor to understand whether there is an effective commitment to support the system and the consistency of allocated resources.

A "good" presentation generally balances aspects of strategic and financial development with the analysis of QS aspects, that the top management takes into consideration for an effective development of their organization. If the first component prevails over the second, the effective support of a QS would be hardly credible.

It is reasonably acceptable that the top management could not have a perfect knowledge of the technicalities and language of QA, but an expert auditor will certainly be able, through appropriate questions, to understand whether quality is within company objectives and at what level of priority.

It is not a minor detail, and essential indeed, the decision made by top management on the level of authority given to QA, and on the appropriate assignments to review the aspects concerning the impact of regulations on the processes of the organization. The decisions that QA must make on the processes and their impact on the organization must necessarily ensure a frank, continuous and naturally constructive discussion, even sometimes in opposition to top management.

During the visit, it is possible for the auditor to observe the true nature of the QS. This can be done through the analysis of the critical points or activities considered generally at risk in the various departments. It is expected that the manufacturing site has made a risk analysis for these critical points, and establish specific, consequent procedures and operating methods in the QS.

In the following sections, some key points are highlighted just starting from the evaluation of the different areas of the site: facility access, production, warehouse, order management, packaging area, and quality control. Obviously, findings of good or "bad" practice can influence auditors and their trust in the consistency of the inspected QS.

Facility Access

In the procedures concerning facility access, it is expected the utmost attention and no "waivers" of registration activities: all the procedures defined for the entry of external personnel must be duly applied and all the relative forms filled and completed, even by the external auditors.

Access control may be complex, especially in PET facilities that are located inside hospitals or research centers; however, it is a critical aspect in terms of product protection as well as radiation protection safety.

The availability of the correct dressing material, often disposable, before entry and the use of radiation dosimeters, which are accurately recorded at entrance and exit, are a good indication of responsibility, knowledge of regulations, and their routinary application at the site.

Production

In the PET world, most of the operational activities take place at night or in the early morning hours. The interview of the working staff during the audit, even if complex due to intense production plans and shift organization, is much appreciated and useful because it allows a direct and broad interlocution.

The possibility of attending productions and actively requesting for clarification, just watching a "live" system, is a facilitation for carrying out the audit and an enrichment for both parties.

The direct dialogue with the operators makes it possible to verify how much they follow the procedures as they have been defined.

Conversely, the opportunity for the auditor to fully understand the processes and their difficulties, which may not always be clearly emerging from just reading documents and flow charts, is largely reduced when the audit is conducted in a plant that is not in operation.

The visit is always accompanied by the verification of the forms available within the departments and used for the registration of data.

The use of updated forms, which are timely completed as prescribed, either during or after the operations, is certainly an indication of good training and good familiarity of the operators with the correct traceability mechanisms defined inside the QS.

A common risk in this area, which the auditor can easily verify, is the presence of uncontrolled copies of procedures or forms, often made autonomously by operators to speed up operations. This "bad" habit can produce copies that almost certainly can escape the control of QA, and also delay or miss updates of the official forms, thus leading to the possible loss of information, use of obsolete documentation, and consequent errors in operations.

The production of sterile drugs (besides radiation protection) requires the adoption of protective clothes, and the gowning procedure is another critical point that is carefully observed by auditors. In this case, for example, especially in the more strictly controlled microbiologically classified area, it is advisable to check how much the change of clothing is reconciled with the necessary and inevitable periods of rest and break time for the staff. Any reuse of clothing, for example, for short periods of work interruption, must in any case be the subject of a risk analysis and strict regulation, with related documented training.

Warehouse

The PET warehouse, given the number of batches produced daily and the limited time frames, is subject to frequent operations of handling of multiple kind of materials.

In this area, the verification during the audit should be focused on a random physical check of the reconciliation of the most frequently used materials, such as reagents, synthesis chemical cassettes, dispensing kits, and primary containers.

For each of them, it should be verified that the tracking of the material moved along the production line is correctly performed by the operators, in real time and systematically through manual or electronic recording systems. In fact, any discrepancy between what is reported on the documents and the materials physically present in the warehouse is easily verified in this way.

Another important point to verify can be the samples taken for analytical and microbiological tests, which are quite in large number and are usually performed by external services.

Some samples could escape the registration and might not be recorded in the electronic or paper system dedicated to the handling of material for internal routine operations, and consequently not correctly entered in the regular flow of controls.

In a regularly operating PET warehouse, with a high frequency of productions, the quantities of raw materials and packaging on the shelves should be commensurate with the production schedules. Otherwise, the use of external warehouses is a mandatory alternative.

During the project phase, the site can hardly evaluate a clear forecast of consumption of materials, because of the impossibility on knowing future evolutions in its production capacity. Consequently, the correct sizing of the warehouse is one of the critical aspects of the PET production.

The lack of space is a common problem of PET facilities; a solution is often found by using external storage services, such as satellite warehouses, in particular dedicated to bulky material like packaging or primary containers.

The effective control of this additional warehouses by the site should be documented by means of contracts, duly defining the responsibilities and procedures shared between the PET site and the external services. The absence of such contracts could be an indication of possible storage points that are not fully under control of the QS.

Order Management and Communication with Customers (Quality of Service)

The PET production plan is structured as a list of customers and of the relative specific quantities of product to be delivered to each of them at a selected time. The list almost always varies for each single batch of product. The order management process and subsequent communications with the customer are an integral part of this production plan.

The order from the customer is forwarded to the PET department through the sales network and customer service, in which it is processed by combining delivery times with estimated production capacity at the selected manufacturing site.

The correspondence between what is ordered by customers and the production plan, in terms of complete satisfaction of requests, is a good indicator of an adequate level of training in which attention has been paid to create an effective connection between the various sectors impacting on the process, a correct passage of information, and, in the end, a prompt response to the customer.

The presence of continuously modified production plans or dispensed doses beyond those requested could be an indicator of an approximate knowledge of local production capacity by the site managers.

It is not a marginal aspect because, if the problem of dose adjustment is frequent, it becomes a source of stress for the entire system which, in the case of PET production, needs an extreme standardization because of the limited time of production and the fast decay of the product.

It is clear that there may be exceptions in PET production, depending on technical and operational variables, but they should not be the rule and must be limited in time and documented in detail.

Packaging and Shipping Department

The shipping department of a PET manufacturing site should be located near the production department and cannot be centralized like in the case of companies that produce radiopharmaceuticals based on longer-lived radionuclides, like those for SPECT.

In both cases, however, the dispatch of the product has particularly important implications and needs of monitoring:

- The control of the leaflets to be included in the packages
- The management of shielded containers and their plastic transport box handling, including maintenance and sanitization

The control of leaflets should not be underestimated, given the severe regulatory implications and the differences between the regulations in the various countries, when product destination is abroad.

The correct management of leaflets and their update is an indicator of the effective and constant application of the approved procedures and therefore of the existence of a functional and robust QS. A useful check to do is the control of the reconciliation between the number of printed leaflets and the quantity of them actually used.

The leaflets included in the packages, being a precise requirement included in the MAA, are subject to revision. Since in a PET facility, leaflets are frequently printed directly at the site, in a certain quantity on the base of the production plans, it is necessary to keep track of them constantly and have a precise evidence of the available quantity, the quantity effectively used and that remaining on stock.

Keeping good track of this numbers, in case of a regulatory change, can allow the documented evidence of the correct disposal of the obsolete copies.

Even the management of shielded containers and the transport boxes could be considered a point of risk, and as such its correct management can be an indicator of the application of the defined procedures. Containers management should include

- The correct flow of incoming and outgoing containers, with demonstration of inspection and cleaning prior of reusing of each item;
- Effective sanitization of shielded containers, given their use in the proximity of the product dispensing station.

The incorrect management of shielded containers may have consequences on product quality, and should not be handled and stocked without an adequate identification of their condition: packaging in/packaging out, unchecked/passed/under maintenance.

Obviously, the handling of these kind of packaging must be recorded in specific documentation, that must be available for any verification.

Quality Control

The peculiarity of PET quality control is to work in close collaboration with the production department for the analysis of samples and the release of batches.

To optimize the shipping times in the case of sequential aseptic dispensing, the samples for QC are the first vial prepared in the process and readily transferred to the QC laboratory before the completion of the vials filling plan.

Consequently, a particularly important aspect to verify is the maintenance of the calibration status of the equipment and this is directly related to the continuous and correct application of procedures. During the audit, it is therefore advisable to collect evidence that all equipment has been checked before being used. To carry out this type of verification, the shipping times and the intervals between productions allow to trace the execution times of the instrument control activities. These feedbacks constitute a valid help to understand if the QC activities have been carried out correctly and if the personnel actually had the time to complete all the planned activities.

Finally, the indicator of a QS in which communications work adequately is certainly the efficient and correct passage of information and data; indeed, also a good indicator of a correct release of the product. In fact, this communication flow includes the documented review of all data, as well as microbiological and environmental aspects, performed in real time and in the correct sequence by all the competent functions: production manager, QC manager, QP, and QA.

If all the above points were satisfactory verified, we can reasonably be "sure" that the system is really in place and works in a straightforward way: it's not a movie!

Finally, we can provide some conclusive considerations.

- An ideal QS, common to all companies, does not exist but it is defined and developed inside any single company. A QS, that is simply imported from another context and applied as it comes, would be a "fake," a "house of cards," not sized for the company and above all not sustainable.

- The development of a customized QS inside the company is always based on the risk analysis performed on all processes; the adequate procedures, technologies, and resources need to be defined to deal with the emerging risks, keeping always their number at an acceptable level.
- A structured QS that reflects the company technology and professionalism is unique, inimitable, and real.

Chapter 6
The PET Manufacturing Experience in a Public Department, a Witness

Contents

6.1 Reasons of a Decision: Developing a PET Radiopharmacy for Internal Use in a Public Structure

Being on the Edge: A Star Was Born

The challenge of the exploration of disease mechanisms and the search for healing solutions have fascinated a multitude of researchers and scholars. Over the time, technology and innovation have pushed forward the number of disciplines that takes part to this quest, have cross-fertilized the field of medicine, and eventually evolved into it, opening unexplored roads and discovering new targets of application. Biomedical engineering is one of this.

Nowadays, the historical role of the family doctor has been integrated by a number of specialists and professionals, operating in technologically advanced medical environments that have grown and spread into hospitals and medical centers. However, it is surprising to notice that a new biomedical technology, even if known for a long time, can suddenly undergo a quick change in fortune and become a blockbuster. Of course, the development of a sufficiently engineered product is a critical gate, but additional conditions play also a role: a wide impact on patients, the managing of costs and sustainability, the easiness of use, and—last but not least—the acceptance by the medical community.

© The Author(s), under exclusive license to Springer Nature
Switzerland AG 2023
A. Pecorale et al., *PET Radiopharmaceutical Business*,
https://doi.org/10.1007/978-3-031-51908-6_6

The newborn technology has often crossed troubled water, and its developers have experienced failures, hypes, and successes. However, when the new technology finds its place in the landscape of the medical field, beside advancing the care of patients, it generates a cascade of new market segments.

This history fully adapts to Positron Emission Tomography (PET). The physical principle (coincidence detection) was first used as a new imaging modality at the Boston hospital in the early 1950s. The PET image represented the distribution of a radioactive tracer in the body, obviously representing the bunch of physiological or biological processes acting of the tracer itself. Radioactive fluoride (i.e., the positron emitter fluorine-18) gave beautiful bone images due to its "affinity" for bone tissue. The equipment was very primitive, and images were just a proof-of-principle rather than something suitable for the clinics. Nevertheless, the method had very high sensitivity and could yield quantitative radiological data: too appealing to be dropped.

Anyway, it was a headache to many physicists to keep the expensive research on PET alive. During the following two decades, biomedical imaging made a leap forward with the newly emerged concept of tomographic imaging reconstruction (discover dates back to 1967), by which radiological images could overcome the limitation of bidimensionality (2D) and look at organs in their trimensionality (3D), and reslicing the images in better resolved 2D post-processed images. A fantastic leap forward! In a short time, complex and expensive imaging instruments, i.e., the computed tomography scanners, became necessary diagnostic tools in the clinical routine. "Seeing is believing" and "seek and treat" are a good iconic presentation of the new impact of imaging in medical or biomedical imaging.

In the end of 1970s, the PET scanner (a bunch of specialized detectors) reached its "ring" configuration, likewise CT scanners, and tomographic imaging became easier and fully available also for this new imaging tool.

At the time, PET was felt as an expensive research "toy" but also a prodigy of technology, promising a kaleidoscope of discoveries in biomedicine. For the first time, the behavior and fate of a molecule could be followed in vivo, and its regional concentration truly measured over the time in a non-invasive way. PET was acknowledged by many biomedical scientists and medical doctors as the new frontier of disease exploration and fight.

Short-lived positron-emitting radionuclides, to be chemically transformed into biologically active PET radiotracers, were the key players for this purpose. Unfortunately, they needed to be artificially produced readily before use, because of their short half-life, and thus closely to the PET scanner, for obvious logistics. So, to go this route, a very complicate and expensive but necessary marriage had to be celebrated between the PET scanner and the cyclotron, a nuclear accelerator capable of PET radionuclides generation by nuclear reactions. Many large research centers across Europe started their programs in consequence.

At the beginning of the 1980s, the scanners (and cyclotrons) were fewer than the fingers of a hand, with the oldest installation being that at the Hammersmith Hospital in London. In Italy, the National Research Council (CNR) started a program as well, with a project of establishing PET centers in Naples (a scanner[1] for neurological applications), in Pisa (a whole-body scanner and a 16 MeV two-particles cyclotron) and in Milan (a whole-body scanner and a 42 MeV multiparticle cyclotron, able to produce both PET and SPECT radionuclides).

Supported by a very positive trend, the industrial manufacturing of PET scanners and dedicated cyclotrons was quickly moving forward, mostly in the US, Europe, and Japan. Indeed, there was a steep development of technologies during the decade between 1985 and 1995, so that most of the centers that could take the challenge were receiving prototypes. Technology obsolescence was so rapid to create troubles to many customers, at that time.

Only the research institutions or the largest research hospitals could afford the new PET technology, and in general this decision was supported by ad hoc grants, which were justified by the achievement of scientific goals. The drive was to be at the edge of medical technologies and has the floor for establishing new discoveries and development, new frontiers to fight diseases.

This was the ground in which PET, during the decades from 1980 to 2000, experienced its transition from research tool to widespread clinical device; it was also the hard ground on which many radiotracers were conceived and tested as radiopharmaceuticals in clinical trials. However, only few of them have eventually reached the clinics.

A Personal Story: An Old-Time PET Meeting

In 1981, I attended my first meeting on PET; it was also my first time at a meeting abroad, and my first time in the US as well. It was held right in the middle of the Big Apple, at Bellevue Hospital, 2nd Avenue, New York. I received the invitation as a visiting scholar at Brookhaven National Lab (BNL), one of the earliest poles in which PET had been developed since the beginning.

PET was the reason I was at BNL: to be primed on the technology and bring home the basic know-how and skills to open and run a PET/Cyclotron site within the project, to start a PET research network in Italy, which was promoted by the Italian National Research Council (CNR).

Only a couple of years before, the PET image of the "thinking brain" went around the world. It was obtained by showing the regional glucose consumption by brain structures following sensory stimulation: music or speech. It was highlighted by a radiolabeled tracer, [F-18]-fluorodeoxyglucose, mimicking the glucose fueling neurons. I was listening to these pioneer scientists and learning their newer achievements.

---

[1]At the beginning, the cyclotron was not installed at this site. An experimental radionuclide generator was used, in which the radionuclide gallium-68 was generated by decay of germanium-68 and chemically separated from the parent radionuclide. The $^{68}Ge/^{68}Ga$ still exists and is widely used in modern PET facilities.

The meeting involved scientists and physicians; they were coming from different main US cities and universities. Among the others, Los Angeles (UCLA), S. Louis (Washington University), and New York (Brookhaven Nat'l Lab.) emerged. They were three US locations with a PET scanner installed.

Beside the need of training further on English (and American English dialects), I came back from the meeting with many reasons for reflection:

- The major players had different views: scientists were asking to have wider access to patients to test new ideas while physicians were asking to have more data on clinical results before enrolling more subjects on new protocols (quite a snake biting his tail)
- The PET scanners in operation were all home-made and results (noting to compare with actual scans) were still largely technology dependent
- The clinical research perspective, even if limited to the brain at that moment, was extraordinary: in principle, any biologically active molecule could be labeled and its fate non-invasively followed in the intact, living patient or normal subject. This meant the possibility of studying in vivo pharmacology: receptor ligands, neurotransmitters, energy substrates, enzymes, the entire brain biochemistry. Or, at least, we thought so
- What was done on the brain could be extended to the whole body, provided the scanners were appropriately modified (and this was already in the mind of Companies based on the US, Europe, and Japan, which were following the PET projects and seeing the new worldwide business).

My mind and those of my colleagues in the lab were full with hypotheses and potential research targets. The horizon was incredibly vast and bright for PET.

The Cyclotron Fever and the PET Tracers

The hurdles of PET are mostly hinged on the short-lived nature of the radionuclide needed to prepare the tracers, from 2 min of oxygen-15 to 109 min of fluorine-18. Therefore, the installation of the cyclotron and its accompanying laboratory of radiochemistry and radiopharmacy were a must have for all those who wanted to start a PET center.

Unmet clinical needs are a powerful trigger once a possible solution is identified. Therefore, following the positive trend of PET perspective, more than 30 medical cyclotrons were installed in Italy in less than 15 years. A similar situation spread almost anywhere; and tens of cyclotrons were installed and planned in a short time, driven by the exceptional impact obtained by the PET exams in oncological patients (specifically, PET/CT from 2000 on). The online database of the International Atomic Energy Agency (IAEA) lists about 1500 installations, most of them are (were) used for PET radionuclides.

The biomedical industry made a large effort to provide the customers with tailored machines that could fit their specific needs. In fact, a "PET customer" could make a choice between cyclotrons accelerating up to four different particles and different maximum energies (both parameters can lead to the production of a broader number of radionuclides), having different beam intensity (thus increasing production capacity), or single particle and low-energy machines, so that even hospitals with a low-budget could afford one.

Indeed, there was a sort of PET/cyclotron fever. In practice, the cyclotron market was saturated in just a couple of decades, despite the complexity and costs of such a solution. In fact, the cyclotron required a heavy shielding (either a room, a vault, or a surrounding barrier, tailored onto the machine itself), and the operational logistics of the PET site included a technologically advanced laboratory of radiochemistry, better known as radiopharmaceutical chemistry or radiopharmacy, between the cyclotron and the scanner. The initial general fever (any tracer for any possible application) was then followed by the consequences of the impact with the real world, i.e., with clinical applications.

In fact, oxygen-15 had a too short half-life to survive a wide clinical utilization, and its application (85% neurology/15% cardiology) was soon replaced by Magnetic Resonance Imaging (MRI). At present, PET scanning with this radionuclide has been abandoned almost anywhere.

Nitrogen-13, with 10 min half-life, requires that the PET scan is performed almost in-line with the synthesis of the tracer. Therefore, a fully dedicated time of the cyclotron, the radiopharmacy team and the labs was due for all the time needed to study the patient (60–90 min). The regional myocardial blood flow can be quantitatively measured with PET and ^{13}N-ammonia much better than with SPECT, but this measurement is very complex and clinically substantial in a very limited number of patients. In practice, apart from research protocols, conventional nuclear medicine could do the job!

The utilization of carbon-11 survives in a limited number of centers, which have developed specific applications in patients. This radionuclide is not prone to industrial development (third-party production and distribution) and remains within the limits of radiopharmaceutical compounding. In fact, no marketing authorizations have been issued for ^{11}C-labeled compounds in Europe. The production is performed for local use, either following relevant pharmacopoeia monographs (e.g., ^{11}C-choline, ^{11}C-methionine) or conducted on the basis of authorized clinical trials.

Fluorine-18 is the most convenient radionuclide from the practical point of view. It has a short half-life (about 110 min) that is anyway long enough to allow more (or longer) chemical transformations, and hence a larger span of uses and a wider availability of product types. However, among the many "radiofluorinated" compounds, few products reached out true clinical applications and 2-[F-18]fluoro-2-deoxy-D-glucose, as everyone knows it better FDG, is at the forefront of the group. As a radiofluorinated analogue of glucose, it can accurately differentiate healthy from neoplastic, proliferating tissue. In practice, 90% of the myriad of cyclotron sites ended up only producing FDG, and, as a matter of fact, the development of the industrial production and distribution of FDG cured the cyclotron fever and sent many of them to an early retirement in many countries.

A Personal Story: The CYPRIS 325 Cyclotron

It would be unfair I missed to include the early history of the "medical cyclotron." Before becoming a truly commercial machine, that can be specifically tailored to the mission a customer wishes to pursue, the cyclotron was directly inherited from the world of physics. In the early 1980s, there were very few companies in this field, and most of them were producing components based on physicists' design more than machines dedicated to the market. However, when the market calls somebody answers.

My first cyclotron was also the first Italian medical cyclotron (i.e., outside a physics department). It was a cyclotron (Cypris 325) manufactured at the beginning of the 1980s by the French Company CGR-MeV, not anymore on the market.

The machine itself was a prototype, developed in a joint venture with Sumitomo Heavy Industries (Japan). CGR-MeV had already produced higher energy machines dedicated to the production of medical radionuclides for SPECT (e.g., iodine-123, tallium-201, gallium-67). The emerging PET business gave them good hope to launch on the market a lower energy machine, suitable for the specific production of PET radionuclides.

The project was based on the ability to accelerate protons to 16 MeV and deuterons to 8 MeV, with a beam intensity of about 50 µA. Quite similar specifications were adopted by the other competing companies, such as the Cyclotron Corporation (US), Scanditronix (Sweden), and Japan Steel Works.

The installation was challenging: first, we had to overcome the authorization of the installation of a "nuclear plant" (the category in which the law on radiation protection located the accelerators other than radiotherapy machines). Then, we struggled to build the underground bunker next to the Institute, which was in the core of the historical Pisa hospital. The soil had the same characteristics to that underlying the Pisa Leaning Tower, which was few hundred meters away. We did not want a leaning cyclotron, nor causing the leaning tower to fall. Fitting the cyclotron and the radiochemistry lab inside the Institute was expensive and complex: the underground bunker had one sole position possible; so it was for the lab, necessarily located at the ground floor (to accommodate the heavy hot cells) and close enough to the bunker to make connections possible. A lot of people at the institute hated me after their activity was relocated.

In 1986, the site in Pisa was officially inaugurated by the national authorities. The cyclotron was initially celebrated by the media; then, a white and black situation followed, because the exciting medical perspectives presented by some media were associated by others to the ghosts of a nuclear disaster (Chernobyl disaster was only a few months earlier!).

The machine was very "home-made," like many devices that you may find in physics lab. Very nice components were machined out of copper and aluminum, a bit Jules Verne's style. My old cyclotron was solely a source of accelerated particles. Nothing similar to the modern "turn-the-key" machines, that come equipped with any sort of optional, from refined tuning of performances and safety controls on the machine, to any sort of automated targetry and radiochemistry necessary to manage and work radionuclides and syntheses. In particular, the targetry had to be developed, i.e., the high-tech containers in which the accelerated particles produce the radionuclides (nuclear reactions) by impinging on specific target compounds (e.g., fluorine-18 is produced by the bombardment of enriched [O-18]-water with protons).

Dealing with the cyclotron, beside the synthesis of FDG I had learnt in the US, quickly made me realize how complicate the road I had chosen was. The target could fail, the beam could fail, the bombardment could fail, and the many cyclotron power supplies each could fail. Many additional variables added to the weak equilibrium of the FDG radiochemical process. I also had to further develop the interface with the medical doctor community. The doctor was not really interested in my troubles, because he had already to deal with the patient, and that was his priority and major concern.

In those years, oncology applications had not exploded yet. In Pisa, PET with [N-13]-ammonia was mostly used in selected cardiovascular patients, and exams scheduled had nothing to do with the numbers I met later, luckily with a new cyclotron and an upgraded and more efficient chemistry hardware.

Still the hanging blade of something wrong from the cyclotron and the synthesis will remain on any future production.

The PET Work-Horse: FDG

When I turn my mind back to the early 1980s, I remember the excitation on the wide latitude of molecules that could be labeled and studied in vivo with PET. Indeed, thousands of molecules have been labeled with the four "physiological" radionuclides,[2] about 4000! However, only a fraction of them have been effectively tested in humans (often few tens of subjects or patients), and, even less, used in controlled clinical trials. Most of them stopped at non-clinical tests (on animal) stage or just at the radiochemistry demonstration level.

It is evident that there is a gap between what is scrutinized by academic research and what can actually reach the patient. The spirit moving Academia is mainly oriented to data publication, possibly on highly ranked scientific journals. This more basic research rarely includes the filing of a patent, which may delay publication and is mostly chosen when a priority tag over an idea is needed.

However, even published results that include early phase clinical trials, albeit responding to all ethical and safety requirements, have no evidential value in view of clinical applications or product development, and remain indicative unless the

---

[2] Oxygen-15, nitrogen-13, carbon-11 and fluorine-18 are called "physiological tracers". They are almost 100% positron emitters, and are available at low-energy medical cyclotrons. They are so called because they are the radioisotopes of the atoms that make up the biological matter (C, N, O), while the small size of the fluorine atom can mimic hydrogen in some instances.

study is further developed. Each of this step is a valuable effort and a worthwhile contribution, but it is not sufficient for a nuclear medicine doctor to take the responsibility to use them in patients outside a research protocol. Eventually, an official controlled study (at least as phase 2) needs to be repeated following locally enforced rules and obligations.

Since the early beginning of PET, the 110 min half-life of fluorine-18 attracted more than the other positron-emitting radionuclide with a shorter half-life. The fluorine-18 half-life ensured longer time operation, multiple chemical reactions and manipulation, and even more noteworthy, it allowed the shipment of samples and dosage vials to reasonably far locations, including satellite PET scanners. As a matter of fact, FDG was the first tracer traveling between the early cyclotrons and PET scanners, and solving site complications to many clinical centers.

FDG had an early penetration and a quickly increasing impact in the scientific literature because, beyond the brain and neurological and neuropsychiatric applications, tracing glucose metabolism is of paramount importance in many districts and pathologies of our body. Therefore, FDG became a target compound for most of the early PET laboratories and, as a result, also the subject of a formidable collective research effort.

FDG was born in 1979 as a candidate tracer to gauge regional energy metabolism non-invasively in the brain. Prof. Sokoloff, from the US Laboratory of Cerebral Metabolism of the National Institute of Mental Health, in Bethesda, had developed a mathematical model to describe the glucose metabolism in the brain. The quantitative data (regional concentrations versus time) obtained with the PET scan of the brain could be fed into the model and kinetic constants of the metabolic process derived. The brain activity could be regionally measured and associated with pathologies, such as dementia progression, or physiological studies, such as brain region activation due to the performance of a task.

A lot of bright neurological clinical research; however, the primary clinical application of FDG would be in oncology. The high metabolic rate of glucose consumption, which is present in the neoplastic tissue, was a fingerprint of the disease, even at the earliest stage of development, when neither radiology nor other diagnostic tools were able to give a response. Furthermore, the signal was so sensitive that it could provide staging and disease grading, and prompt indication of the response to treatment as well.

FDG caused a major change in the way the oncologic patient was managed and the exam became necessary in almost any cancer patient.

However, the number of exams still remained low for quite a long time; they were mostly obtained in clinical trials and research applications and aimed at demonstrating the clinical impact of PET in the multiple manifestations of cancer. Initially, PET was struggling to find its way in the hospitals and the PET scanner was often regarded as a complex and too impacting device, due to its associated harness. On the other hand, the applications in oncology (cancer detection and staging) were becoming more and more relevant and efforts were made to rationalize the costs, at least for what concerned the availability of FDG without the economic impact of the cyclotron-radiochemistry costs. The United States, who were leading and strongly pushing the clinical development of PET, was first to establish and operate a network for FDG distribution (1996). It was based on the pre-existent regulation on centralized radiopharmacies, which were already active in support of SPECT diagnostic services. However, the real turning point of PET fortune would come shortly thereafter.

In 2000, the development of the hybrid imaging modality, i.e., a whole-body PET scanner coupled to a computed tomography machine (CT), paved the way to a breakthrough use of this diagnostic approach by which, for the first time, metabolism (biochemistry) and anatomy could be assessed in a one-stop-shop approach. FDG had already shown the ability to trace tumor lesions, at a very early stage and with quantitative measurements, which were able to determine disease staging. By adding the anatomical data to the metabolic information, the doctor had full definition of the lesion and, equally important, a quick indication of its response to therapy and treatment. The CT protocols based on morphological changes could be integrated with metabolic response to treatment. The impact was that response could be often evaluated in weeks instead of months, with great benefits on patient management and effectiveness of treatments. The acceptance by the medical doctor community was enthusiastic and the PET/CT exam with FDG deemed as mandatory in cancer patients.[3]

PET/CT is not the only booster, a strong research effort had also concentrated on the production of FDG and its fluorine-18 parent radionuclide. In 1986, a breakthrough method of synthesis was developed, so that it was adopted worldwide. In the following decade, there was a formidable improvement in the yield of FDG radiosynthesis, as well as in the efficiency of the cyclotron production of fluorine-18. The production of FDG was readily feasible, adapted to run onto an automated hardware, and eventually suitable to large-scale productions.

In principle, FDG could take the challenge of a heavy duty cycle. There was one more, last point to address: the sustainability of PET/CT exam, despite its costly nature (a PET/CT scanner is worth several M€ investment). The increasing impact of PET/CT in oncology as an imaging tool, quickly highlighted relevant benefits: improved patient management by sparing unnecessary/detrimental treatments and nonnegligible economical savings. This information found its way in medical literature and reached a large visibility between professionals and lay people too. As a consequence, the pressure increased on having PET exams admitted to reimbursement by the Insurance Companies or the National Healthcare System (where this exists). The flow of money soon became relevant.

In this new scenario of a high social demand of upgraded technologies, it was highlighted that the production capacity of a single FDG/cyclotron plant was high and could easily serve more than one scanner, even placed at different locations. The rationale for a new organization became evident: roughly, a cyclotron facility could prepare large batches of FDG to be split and distributed to a number of

---

[3] As a matter of fact, FDG was the pillar of PET/CT fortune and vice versa. In 2020, more than 85% of PET/CT scans were made by using FDG as the tracer, and diagnostic routine only limited by FDG availability.

satellite PET/CT scanners, located in a radius of few hundred kilometers, or better within a 4–5 h driving time (door-to-door considering the radionuclide half-life of about 2 h). Each hospital could place an order based on its schedule of patients.

The old formula one-cyclotron/one PET center was under discussion in many situations, whether economically obsolete and not viable. The new paradigm of FDG delivery started to spread worldwide and the idea of a business associated with its production as well.

The projects immediately faced some hurdles, i.e., coping with the applicable rules concerning the distribution of a short-lived radiopharmaceuticals (drug). Historically, FDG had been mainly used on-site, in hospitalized patients, referred outpatients, or in clinical research subjects.

The western world was firstly developing the FDG business, and in practice, it divided in two regulatory areas, on the basis of the enforced regulations issued by the Food and Drug Administration (FDA) or the European Medicine Agency (EMA). FDA and EMA had different rules and regulations on radiolabeled drugs, site licensing, product registration, and distribution. Furthermore, Northern America and Europe had a different history on the use and approval of radiopharmaceuticals as drugs, and radiopharmacies as manufacturing site. FDA had more practical and clear rules, as well as an ongoing experience from the control of centralized radio-pharmacies, which have long operated in the US for SPECT dosage distribution.

There was a strong and generalized sentiment of the need to have a centralized FDG distribution, and the option to install (or keep) an in-house cyclotron was increasingly seen as a very delicate decision (costly, complex, discontinuous). Meanwhile, those centers who had larger competences to run FDG productions and had suitable technologies, pondered the same questions: should I take the chance to start a project aimed at FDG delivery? What kind of a business case should I choose? Will my laboratory/organization endure the effort? How can I get the money to start? How can I solve the obligations imposed by the drug regulations and pass the inspection by the competent authority?

The center of Pisa was in a similar situation, and these were the questions I asked myself.

A Personal Story: The Synthesis of FDG

I was very excited when I received the clearance to attend my first synthesis of FDG at BNL, in 1981. I had already seen the lab: the entire synthesis was manually performed by two operators. The equipment was placed behind a lead wall to protect the operators from being exposed to radiations, a small lead glass was placed in front of the reactor (a modified Rotavapor® used to remove solvents during organic chemistry reaction processing); small tubings connected the reactor with ad hoc glassware and a final recovery ampoule. Works were in progress to install a brand new hot-cell with a large front lead glass and a master-slave manipulator, so that a larger activity of molecular fluorine-18 could be handled safely and more efficiently.

I knew, from reading the process outline, the synthesis would last about 2 h since the trapping of the radiolabeled gaseous precursor, the very aggressive ^{18}F-F$_2$, which was produced at the 26-inches cyclotron. The expected maximum yield of the process was around 10%, including FDG purification.

I saw with surprise the two technicians entering the lab. One was wearing an American Indian headdress (the dosimeter was at the front center, I realized later on it was so because they needed to lean out of the lead to watch a part of the harness), the colleague had a long, silky blue scarf winded to the neck. I could not refrain and asked why they had that garment. The answer was unexpected: because otherwise the synthesis may fail!

They may have tricked me, but later on I recognized how variable was the outcome of the production, and the occurrence of failures.

There is a basic fragility in the synthesis of FDG: the tiny amount of product that is treated (well below micrograms). The precursor fluorine-18 has to be incorporated into a substrate molecule and all reagents, chemicals, and solvents overwhelm it.

I could witness it practically. Once the synthesis of FDG failed, no product could be flown to the University of Pennsylvania. The synthesis keeps failing, no ways to get again the product, and the rented touristic airplane kept waiting at the Islip Airport. It took days before we could find the culprit, and it was quite unexpected. A bottle of aluminum oxide, used in the purification step at the lab since the beginning of the synthesis (it was a big one), was replaced by a new, identical one; same brand, same product. No more FDG was obtained until the dry powder reached an equilibrium with the environmental humidity: the purification dynamics was altered in the extra-dry powder. Since then, a check over that critical step of FDG was established, but almost 2 weeks of panic were elapsed.

The following adoption of remotely controlled synthesizers, for a long time developed in-house, reduced timing and operator errors, which are always ambushed in a whole chemical and pharmaceutical process that is compressed in less than 2 h.

Few years later, I started my lab in Pisa. Over the time, I went through the different technologies that, year after year, have been developed to give more stability and reliability to the process. Even a new cyclotron replaced the old CGR-MeV prototype. However, the maximum effort was on radiochemistry, the "cooking lab," in which a chain of technical solutions have been tested over the time.

After initially employing classic DIY automation (do-it-yourself), it appeared that employing a robotic arm would give the best results in the synthesis of FDG and, potentially, other tracers as well. The robotic arm moved back and forth along a rail on the back wall of the hot-cell and was surrounded by a circle of "intelligent" stations (weighing, mixing, heating, distilling, purifying, diluting). However, it turned out to be a labor-intensive system, because the robot had to be continuously controlled, to prevent self-injuring intentions of crashing against the side walls of the hot-cell (the bug in the software was hard to find and fix!). Therefore, it was replaced by a new automatic module, based on microconnectors and minute section plumbing to reduce the volume of the system. The hardware had to be assembled by the operator, hand-loaded with reagents and consumables, and remotely started, so that a time-list guided the process. Nevertheless, operators' errors (e.g., incorrect reagent loading or wrong connection of components) mixed to intrinsic, ghost synthesis failures. The latest update was the use of pre-assembled cassettes, with connections already set up and pre-loaded with all reagents, which could be plugged into an automated hardware. This solution lowered operator's mistaken operations and could even benefit from manual corrective actions made in-process by the operator in case of need. A list of dedicated plug-in cassettes was commercially available and allowed the preparation of radiopharmaceuticals other than FDG, just using the same hardware, thus widening the latitude of tracer available at the site. The step forward was significant.

During this long experience, the original FDG process has so deeply changed since my initial US experience. The use of molecular [F-18]-F_2, a very aggressive and instable labeling precursor, was replaced by [F-18]-fluoride and the electrophilic synthesis by the more efficient nucleophilic process. The chemical hardware is beyond comparison with the previous one. Nowadays, the time of synthesis has dropped to about 30 min and the yield increased to 60%.

Still, a limited but existing number of failures remains physiologically embedded into the FDG process.

Setting an In-House PET Radiopharmacy Up: A Long History

There is a pioneering impulse to be the first to test a new technology in medicine, and this has often sparked interest of large research hospitals to open PET centers and offer the most advanced diagnostic tools to patients. This has contributed to the spread of cyclotrons and radiochemistry laboratories that burgeoned worldwide. Unfortunately, many of these projects responded primarily to hospital marketing motivations, such as being characterized by a state-of-the-art offering of diagnostic modalities, rather than a pondered decision that considered the underlying level of cost and complexity. As a matter of fact, the start of a cyclotron and its radiochemistry (radiopharmaceutical) laboratory was often underestimated.

Some fascination came also from the chimera of the push-a-button chemistry for the synthesis of radiotracers, which was presented at nuclear medicine conferences by many manufacturers of synthetic modules. In the past, most of these early automatic devices were mainly spin-off products, placed on the market by scientists and engineers who had developed them as prototypes during their research and often could not survive what turned out to be a very volatile market. Good products died because of insufficient technical assistance or market demand.

The cyclotron business experienced a golden age, driven by a steep growing of the number of installations, in which models and manufacturers have flourished and died. This was the top moment for the spreading of in-house PET radiopharmacies. Even general hospitals, provided they had a nuclear medicine department, rushed to have a cyclotron to feed its scanner(s).

The point of survival was in the background expertise and the basic know-how existing or being created at the site. Two points were particularly misconsidered: the costs of managing such a complex infrastructure and the productivity and skill of the radiochemistry personnel.

In general, the oldest and largest radiopharmacies were able to withstand the critical phases, such as the selection or the change of a technology and the ability to master the basic technical and scientific problems associated with the effort.

The idea of starting a radiopharmacy at my institute was established long time ago, around the 1950s. It was in line with the emersion of nuclear medicine as a discipline in Europe and the establishment of the earliest nuclear reactor plants dedicated to medical radionuclide production. The bridge established between the institute and the newborn European healthcare industry, ensured a continuous exchange of research, technology, and practical experiences aimed at the development and validation of in vivo and in vitro diagnostics. Indeed, many early products in the field of radioimmunoassay and SPECT radiopharmaceuticals were developed also with the partnership of my institute.

In 1980, when I took my position at the Institute, the radiopharmacy was fully operational. Commercially available radiopharmaceuticals were daily used for the in-house clinical needs and routine operation included the preparation of classic technetium-99m labeled kits, the formulation of ready-for-use radiotracers ([Tl-201]-thallium chloride, [Ga-67]-gallium citrate, [I-123]-iodinated tracers, etc.). There was also a well-developed basic research on new tracers, mostly labeled with Tc-99^m and I-125 and I-131. What surprised me was the number of chemists, physicists, and engineers working at this medical research Institute. There was a wide availability of unexpected basic science resources, from informatics to analytical chemistry and biochemistry; even a machine shop, in which one could have his prototype designed and constructed! A lot of fun for a young scientist in charge for starting a PET/Cyclotron facility.

At that time, the situation of radiopharmacies and radiopharmaceuticals was different, in Italy and in other European countries as well. Nuclear medicine diagnostic products on the market comprised SPECT diagnostics and radioactive therapy capsules or solutions. However, no product used in the clinics had to have a medicinal product license throughout Europe until 1989, when a specific EU directive was approved. Since the establishment of nuclear medicine (in hospitals or clinical centers, and doctor's office in some country as well), each European Country (and involved parties therein) established national criteria, at different level of official recognition, to manage the practical aspects of radiopharmaceutical preparation and use. Only few countries in Europe took an early decision to approach radiopharmacy from the pharmaceutical point of view. More frequently, and that was the situation in Italy, the applied regulations concerned personnel and environment radiation protection. In daily practice, apart from the use of shieldings and the adoption of radioactive contamination protective tools derived from an EURATOM directive, routine preparation of technetium-labeled kits and radiopharmaceutical dispensing were assigned the same practical handling rules of common drugs in the wards, such as injectable antibiotics. In many countries (Italy), the radiopharmacy consisted in a small dedicated space, usually as small as possible, in which a table or a fume hood hosted local barriers to shield the operator (a nurse, a radiographer, or the same nuclear physician) while handling the radioactivity. A wall of lead bricks was often used to protect from the Mo-99/Tc-99m generators, generally regarded as the bad guys of the area. A shielded safe, with lead drawers, was used to keep on storage the longer half-lives radionuclides, such as Iodine-131 capsules, and ready-for-use products, such as Thallium-201 and Gallium-67 labeled injections.

The manager of the situation was the health physicist (or the qualified expert) in most of the cases, and the prime target was to ensure radiation protection and avoid personnel exposure and contamination.

The products mostly required very simple manipulations: (a) withdrawing a radioactive solution from a bulk vial; (b) transferring a measured aliquot of radioactive solution inside a vial containing a freeze-dried cold precursor (depending on the target of the SPECT scan); (c) measurements of the result (the activity inside the vial and/or the syringe). Operator's skill and experience were very important to keep hand exposure low. Sometime the vial was to be gently heated before dispensing, or a delay time to be adopted. Much simpler operations than the PET chemistry. Quick quality controls were described by the manufacturer in the instruction for use that accompanied the cold kit. They were in the form of rapid tests, mostly aimed at the verification that the kit was correctly reconstituted, and make sure the nuclear medicine doctor, who held the responsibility of the injection, that the patient would receive an effective injection and scan. A defective preparation could alter the distribution of the tracer and cause the patient an unjustified exposure.

Interestingly, there were official references for the quality control of most of the radiopharmaceutical products available on the market; they were described in monographs of the European Pharmacopoeia,[4] and initially promoted by Nordic European Countries. However, such compendial quality controls were not mandatory in all EU Countries, i.e., prescribed by a health authority. The quick tests were routinely performed and communicated to the nuclear physician so he could take the decision to inject the patient.

The transition to the full application of the newly arrived regulation, i.e., obtaining a MAA for all products remaining on the market, was critical for companies, even if a waving period was established to allow the radiopharmaceutical companies to apply to the Health Authority (HA) for licensing their products. An abridged procedure, which included mutual recognition agreements among EU countries, facilitated the task, but the entire production processes needed to be revised

---

[4] Ph.Eur. methods are "reference methods" and are essential in case of disputes. Alternative methods can be adopted after validation (however they must lead to the same pass/fail result as in the compendial method), and in many cases after health authority approval (however, the competent authority acts on the bases of the body of Official Laws of a Country). The "General Notices" at the beginning of the European Pharmacopoeia is an interesting reading to get oriented in the correct use of all texts (e.g., Monographs, General Methods, General Chapters, etc.) and conventional expressions.

according to the new applicable regulation (GMP) and aligned with the obligations already enforced for medicinal products having a marketing authorization.

A further consequence of the new EU directive, affecting PET radiopharmacy and FDG, was that a clear difference had been established between those who produced for internal use (in-house) and those who aimed at distribution: the latter had to start an industrial production.

The transition of radiopharmaceuticals toward the application of drug regulations triggered a rising attention to quality and safety and an additional effect: it become more and more evident that having a well-organized radiopharmacy could be an advantage for the nuclear medicine practice and an added value for the hospital management.

Meanwhile, the success of PET chemistry and PET tracer development drove an increasing number of non-medical professionals such as chemists, pharmacists, and biologists inside the nuclear medicine community. The pressure coming from the increasing demand of PET scans (oncology and FDG/PET were demonstrating their strong connection), the positive discussion on the reimbursement of PET/CT exams, the considerable spread of PET centers, the involvement of pharmaceutical companies and regulatory laws, would soon synergistically change the face of radiopharmacy and radiopharmaceutical chemistry as a science.

At the beginning of the 1990s, cyclotrons and PET radiochemistry labs were booming at hospitals. It was a quick evolution; therefore, the in-house management of PET production was mostly assigned to the available personnel, in particular to those who had experience with radiopharmaceutical product compounding, when such a resource was available. The personnel was often trained on-the-job, without a specific degree in radiopharmacy (still not existing in many European countries) and without considering that legal limitations started to emerge on who-could-do-what, whereas chemists and pharmacists were not always equivalent, when dealing with drug control and release.

As a matter of fact, the larger the number of patients, the stronger the need to increase batch numbers and dimension; in consequence, the complexity of equipment rose as well, and additional and specific professionalism became prescribed or unavoidable, such as pharmacists, chemists, and physicists, with specific qualifications.

The radiopharmacy—often part of the nuclear medicine department—increased in size, competences, costs, and careers. Sometimes the discussion ended up even in a sort of competition among chemists, pharmacists, and nuclear physicians on the responsibility of specific actions and skills.

Managing a cyclotron and the attached pool of equipment and professionalism, with more rules and constraints, turned out to be a complex and very expensive activity. In particular for those institutions that were working with full-cost accounting, i.e., all personnel costs and additional direct costs—such as cyclotron and equipment maintenance—were on the final bill.

The advent of the distribution of cyclotron radionuclides and PET radiopharmaceuticals (mostly FDG) contributed to a major change in the overall scenario, which has resulted on the one hand in the industrial availability of FDG, with all the practical advantages and safety guarantees of a licensed product, on the other hand on reducing the organizational load for hospitals and clinical centers that have PET/FDG as their main core business. In a short time, outsourcing FDG was cheaper than producing it in house. Furthermore, it was easier to rely on the responsibility of an external manufacturer than face the problem internally in case the FDG synthesis failed and patients missed their doses. As a matter of fact, most of the facilities that were unable to organize very large/multiple productions per day (based on early morning shifts), or keep in their portfolio additional tracers beyond FDG (e.g., nitrogen-13, carbon-11 or radiopharmaceutical used in clinical trials), seriously considered closing their PET radiopharmacy and relying on commercial products.

A Personal Story: Working in a SPECT Radiopharmacy

In 1980, my longer term task was to start the PET radiochemistry at my institute, and this included the installation of the cyclotron. However, as the project progressed, my assignment was in the SPECT radiopharmacy. While studying the theory and waiting to have a full immersion on PET tracers abroad, I began a training-on-the-job on technetium generators and "classic" radiopharmaceuticals. This priming had a number of outcomes: first establishing a safe approach to radiation risk and radiation protection; second, the most important, understanding the relationship between the preparation of a dose and the need of a patient. I had no experience of working in a clinical environment, just into a bad-smelling organic chemistry labs (fume hoods were not as efficient and available as in modern settings).

I was lucky because the SPECT radiopharmacy at the institute was very well organized. The cooperation with the industry definitely played a role in this effort. A full lab was set up, with full equipment, such as gamma and beta counters, and chromatography instruments equipped with radioactive detectors. [Tc-99m]-generators were shielded and hosted within a special fume hood, and labeling of cold kits performed in a protected environment nearby. Different dedicated gloveboxes were used for radioiodination reactions: [I-131]-iodide was used for the labeling of proteins and small molecules (e.g., guanidines and amines), which were used for scintigraphy; while [I-125]-radiolabeled proteins were prepared for the development of new RIA tests or loco-regional therapy of neoplasms with Auger electrons.

After few months, I had enough radiopharmacy practice to enter the clinical protocol staff, and the beginning was somehow shocking. It happened during a protocol on emergency cases following a head trauma. I had to prepare and deliver the technetium-labeled dose to a doctor waiting in the gamma camera. I rushed the dose to him and saw the patient: a young girl, maybe in her 15, laying on the gamma camera bed and waiting to have a lung scan. She was unconscious, though breathing spontaneously, a small bandage covered a head injury. It was a strong emotion for a young chemist never exposed to this sort of a situation, which are so frequent at hospitals. The people, the patients suffer. It took a while to me to elaborate this experience and realize that would have been my job choice: preparing radiopharmaceuticals to patients and not only to the doctors.

A couple of years later, when I came back from the US, there was an internal reorganization and I was appointed head of the radiopharmacy. Thus, I had not only to manage the preparation of the doses, but also ensuring the supply of precursors, generators, and all other tracers used at the institute. Furthermore, I had to organize the staff and keep the relationship with a number of professional figures and officers: radiation protection, procurement office, maintenance, administration, doctors, and nurses. A complex universe that was alien to me while doing dose compounding at the bench, and that made the picture clear to me that radiopharmacy is a chain of events, with responsibilities that interlace and require efficient actions and communication. The patient is at the end of this chain; and this must never be forgotten.

Therefore, my team and I were aware of the responsibility we took on when the cyclotron was installed and the PET radiopharmacy begun to produce, in 1985, and became fully operational the following year.

I was working at a clinical research institute, and PET protocols and tracer production had to be set up almost from scratch. The first radiotracer we produced was [N-13]-ammonia. It was used to study myocardial blood flow in patients with ischemic disease, by using a scanner that was one of the first whole-body PET built by CTI (Knoxville, USA). We worked the project out in-house, in cooperation with industries and international friends and colleagues: but this is a different history, because, at that time, radiopharmaceuticals were not considered drugs, and European legislation still to come.

A Personal Story (Once Again): The Airborne FDG

Once we got the new whole-body scanner, with a larger and longer field-of-view, the push from oncologists was pressing. Routine production of [N-13]-ammonia was running well: good figures were reached in PET imaging of cardiology patients; many of them were scanned for the search of myocardial ischemia and myocardial viability (double scan of [N-13]-ammonia followed by a FDG scan). Furthermore, intense clinical research was conducted at our institute with the aim to validate a new method for the regional quantitation of myocardial blood flow (a software tool subsequently used by many international groups).

However, even if the institute "core business" was cardiology, the pressure of oncologists raised even further and the local nuclear medicine department was not yet equipped with a PET scanner. The synthesis of FDG produced enough activity to cover the "patient slots" of a full day per week of oncology scans. In a short time, 2 days per week were allocated to oncology scans. In few months, cardiology was reduced to 1 day and we were running full engines mostly on FDG, and a waiting list of oncological patients.

One day, the cyclotron was forced to a stop that would have lasted quite long. Too long not to find an alternative solution. At that time, the only FDG manufacturer with a valid MAA in Italy was abroad. It was already delivering FDG to a limited number of scanners located very close to the Italian boundary, but there was no way to have FDG transported to Pisa by road: too far away, too much decay time. Thus, we proposed a solution: flying FDG to the Pisa Airport. I remembered the old time at Brookhaven, when we were flying FDG from Long Island to Pennsylvania University. The proposal was made, and the manufacturer was available to arrange for a delivery by plane. Soon, a price was negotiated for the airborne FDG. There was a positive attitude in this because the manufacturer was also interested into opening a new route to the Italian market, provided we could find a solution on how to have the FDG landing to Pisa.

FDG was a radioactive product, and special rules apply for its transit through an airport facility. Milan, Rome, and Turin airports had all that would have been needed, such as internal procedures on how to disembark and move around the radioactive stuff, trained personnel and dedicated stock and transit areas, but once again the time to reach Pisa from any of those airports by road would have made it unfeasible. The City of Pisa had no authorization to manage the transportation of radioactive stuff across its airport facility.

Then, we invented a solution: FDG would not be crossing the airport facility. We discussed a special, ad hoc procedure with the aviation authority in charge for the airport. The airplane, a turboprop rented by the manufacturer, would reach the general aviation sector of the airport, the same used by the air rescue teams and their medical flights. The airport had a standard procedure for granting access of ambulances and emergency vehicles to such an area. We based ours on this: the local carrier, having a license for the transportation of radioactive materials, would have been waiting outside a special transit gate next to the general aviation area. As soon as the airplane would have stopped at the area and its condition be safe for disembarking FDG, the gate could open, and the transporter enter the general aviation area and position alongside the plane. In that way, FDG could be quickly transferred from the plane to the truck. In few minutes, the FDG would be on its way to the institute, and some 15 min after at our doors; meanwhile the plane was again on the runway, just after refueling. Later, we knew that additional stops had been added by the manufacturer afterwards; in fact, other airports had adopted our procedure, and the manufacturer made its new Italian business more profitable.

6.2   From a Research Lab (or a Clinical Internal Service) to an Industrial Set Up

The Approach to Business of Academy Vs. Industry

Healthcare or "personal well-being"—as nowadays we like to refer to anything is related to heal, correct, improve the everyday life of a patient—is one of the most powerful economical and financial motor, moving people, money and therefore politics (which means the development of legal compliance and financial tools). Health solutions, from holistic to technological approaches, always receive prompt attention by the public and, though with different time frame, by possible actors who can enter the market.

Research institutions are more and more among those players: they can be spurred by the perspective of an economical benefit that can reduce the anxiety of fund quest, and by the will to be active in the promotion of healthcare and societal benefits, such as increased patient access to advanced therapies and diagnostics. In reality, each of them should also answer some fundamental questions: why should I change course from research to business and navigate new and potentially troubled waters? Do I have the necessary equipment? How safe (in a future prospective) and justified (retrospectively) is the economic investment?

In general, there is a foggy horizon before a scientist, when he ponders decisions about the economical value of his own research and technological achievements, and how/whether they should be developed to start a business (and have a revenue). Lifetime scholarship, hours in the lab, international conferences, high level publications on outstanding journals: is it sufficient to establish a decision on "selling" his own experience and expertise? Is the owned "product" (and the associated vision to catch funding) worth the effort? A saying, largely used in Italy, should be kept in mind: "Any baby cockroach is beautiful in its mom's eye!," which means that researchers are deeply in love with their own ideas, discoveries, and expertise, and thus expect that the same feeling is shared by any investor, developer, company, and multinational corporate industry!

On the other hand, research survives on the continuous search for funds, and the need of providing fuel to the lab is vital to any scientist. How less or more complicated this can be is heavily country-dependent. It is eventually linked to the sensitivity matured by politicians on what is, or should be, innovation and science and their ability to make efficient funding tools available to applicants.

In general, there is a very strong competition on funds and the calls are crowded. Meanwhile, the calls are getting more and more strict on the project definition and selection criteria of applicants. Once a project is funded, the funding agencies concentrate on a severe monitoring of task progression and full coherence with milestones and endpoints. Therefore, funds are strictly available for that specified research and cannot be used outside the approved project: for instance, the acquisition of an instrument that is not entirely depreciated within the end of the project. Two problems may arise out of this situation: first of all, funds are not available to the lab in a continuous flow, but follow the frequency of winning applications; secondly, free basic research and technological innovation, such as a high-cost equipment acquisition and instrumentation replacements, may remain unfunded, unless the institution did not develop an independent tool to support it (a difficult decision during time of budgetary shrinking).

In the latter condition, it becomes natural for a scientist to look at the achievements reached in the lab as a resource, and try to put them on the market, pursuing the leap from a short-breath budget (or no budget at all!), to a longer term economical resource, that could finally bring new financial resources to the lab, and help forgetting the frustration of swinging funds and the strict expenditure constraints and reporting rules of grants received from funding agencies.

This is why a research lab may become "willing" to endeavor a risky and unphysiological track. Here, the term "unphysiological" has a specific meaning, because it well describes the stark difference between the way academic research normally develops versus how industrial manufacturing is to be conducted and finalized.

Coming back to PET radiopharmaceuticals, the non-industrial production site may have a different identity, on the basis of the underlying cultural humus and core competencies. It could be established in a research setting, like a chemistry or physics department or a clinical research department, as we have discussed so far, but also embedded in a clinical facility, such as a hospital. Many hospital PET radiopharmacies have been allured by exploiting their production potential to receive a revenue in return. However, whatever be the origin of the site, it will quite certainly have to work hard on re-balancing its core-expertise and organization. For instance, the research center has to understand the importance of patient's needs, clinical routine organization, and timing and maybe healthcare bureaucracy. The hospital may require to develop additional skills and competences within its staff, to keep the production process alive despite chemical and technological failures. Both should identify the point of balance between the effort (or the amount of product) dedicated to in-house needs versus the commercial scope.

Nowadays, the future industrial site can take two different routes: starting the manufacturing of owned (licensed) products or, starting a cooperation with an industrial partner to manufacture the product(s) in its portfolio (third party). The regulations enforced in the country in which the manufacturing site is located can have an impact on this decision. In the US, FDA issued clear guidance for the licensing of both radiopharmaceuticals and manufacturing sites, including those placed inside clinical centers, which may qualify and distribute their products to satellite PET scanners. Indeed, the distribution of radiopharmaceuticals, even in dosage forms, has a long tradition in Northern America and many US PET centers have followed this track.

In Europe, EMA has a different approach: radiopharmaceuticals are treated just as any other medicinal product; therefore, their approval (licensing)—even in a clinical center—comes with the regulatory burden of ordinary drugs, i.e., with all the steps of product development and connected constrains. Quite obviously in Europe, the majority of PET centers who decide to become industrial manufacturing sites either are private entities, or are public institutions and have an industrial partner at their side. Public entities, in which are comprised research institutions and universities, may have limitations to establish commercial/industrial activities, that may be outside their core mission: usually it is health care for hospitals/clinics, and science and technology innovation for research institutes. In practice, they cannot accommodate a project to start a manufacturing site in their budget. Instead, they can initiate projects in which the commercial activity is carried out in a way that is subsidiary to their main mission. Therefore, in a public–private partnership, the first task of the industrial partner is to provide the seed to start the project. The public partner can then participate to the joint partnership with its own resources (with a budget or in-kind). In any case, the project goal must have a high impact on the advancement of technology penetration and social benefits, and not have the characteristics of mere commodity manufacturing.

The partnership is the beginning of an intriguing journey, which includes a wide dynamic of interactions and multiple facets.

A Personal Story: How I Started a Project to Become a Licensed Site

After a number of years of operation of a PET/cyclotron facility, there are only a few reasons that may push you to cross the boundary line of in-house production and move on to distribution: (1) the desire (or the need) to make a step into business, and try to access extra funds to your department, or (2) the response to a political or social pressure that rises because the (local) demand for PET diagnostic is not satisfied by the existing arrangement of resources or patient demand.

In my experience, both were valid. The first because we had moved the entire lab and the new PETtrace® negative-ion cyclotron to a new site, with larger spaces and more efficient layout, and we wished to upgrade the equipment. The second, because we considered as strategic the vision of a site complying with full-GMP regulation and being able to ensure FDG availability to PET centers (geographically at hand), whose number was expected to quickly increase in the near future. In fact, the decision was taken in 2002, and many hospitals had already received a PET/CT, just arrived on the market, or were about to do so; therefore, the need of FDG had to increase.

Meanwhile, the effort of many hospitals to keep alive a full in-house production (including the cyclotron) was both becoming economically difficult and complicated by the need of trained and qualified staff. I had many contacts with medical doctors who had responsibility of nuclear medicine departments: many of them were worried about the costs and the complexity of keeping the cyclotron and the attached radiochemistry. Other countries had started the distribution of FDG, in particular the US, but they could rely on different regulations. Few years before, the entire radiopharmaceutical business was stormed by the obligation to qualify the industrial manufacturing sites, which had to be inspected and get the license for all products in each of the European member state in which they had a commercial interest. In the nuclear medicine community, there was a general expectation that somebody took the decision to face the challenge of licensing a PET radiopharmacy with the health authority. The qualification to be reached was "Officina farmaceutica," i.e., "Pharmaceutical manufacturing site" like any other pharma company.

No previous experience or reference was available in Italy, in particular, no Officina farmaceutica existed outside an industrial setting, and among them, none dealt with PET products at the time. I decided to take the challenge and be the first one. I immediately discarded the idea of going through a registration (licensing) of the FDG that I was producing. It would have been more reasonable finding a partnership with an industry.

During the past years, I had many contacts with the radiopharmaceutical industry, not only as a customer for the SPECT products but also as an expert (or at least an old guy) in the field. Therefore, it was easy for me to explore whether an industrial partner might be interested into developing the site in Pisa as a GMP facility. An interested party was found, and a feasibility analysis started with the partnership of an English Company, who had already the idea to start the FDG distribution in Italy. It owned a proprietary dossier (MAA) on FDG, which was already submitted to the Italian Health Authority, and an advanced project on establishing an Italian GMP site for the manufacturing of such a product. It was interesting, on this regards, the possibility for me to visit a "living" commercial private PET site, and have a clearer idea about the efforts needed to transfer the same operational activity in my institution.

Obviously, as an employee of a public institution, I received the green light on the project by my director. In advance, I had made a focus with him on what could be the partnership contents in terms of returns and benefits, and also the possible development of the project in the future. In other words I defined a "business case," in which I estimated not only the investments on equipment and facility refurbishment, but also the employees to be dedicated to the project and the cyclotron duty time which was necessary to satisfy the commercial demand. This firsthand evaluation was based on the estimate of PET productions planned per year by the industrial partner. Therefore, I had in mind the main points that could enter the negotiation with the industrial partner, and, at that point, a strategy started for a courtship, or at least a sort of. The business target was focused on FDG and not any other PET tracer, because it was already on track to licensing, and its utilization was well accepted by the medical doctor community, which is a "must" to ensure the fortune of a tracer (and a business).

In my earliest experience of negotiation, the agreement with the industrial company included a support for the investment in equipment (cells, synthesis modules, dispensing system). It was based on a 5 years duration, potentially renewable for the same period of time. Finally, the project was simple and straight enough in its headlines to be readily submitted to the internal competent offices (mainly economical and legal) that would be asked to provide an opinion. It did not take too long to have an official start of the cooperation, which was in the form of a confidentiality agreement so that an agile communication could start between the Parties.

Subsequently, when the layout was further expanded and made compatible with more production lines, I paid more attention to some details of the mutual limitations that may be included in the agreement, in particular when the industrial partner supports the public institution with equipment investments. It is important to discuss and agree on the possibility that the public body could use the structure also for activities that can be independent from the contract. For example, the possibility of in-house manufacturing, drug development, or even manufacturing of other partners' products should be maintained, obviously avoiding competing interests between different partners.

Building the Cooperation

The exploratory steps of a public–private partnership are generally driven by the match between the reputation of a public partner and the decision by the industry to invest in the particular business, namely PET tracers. The feasibility of such an alliance clearly depends on many factors, but some of them are quite preliminary: a consolidated experience at the public site (mature technologies and skills), the prospect of economical revenue, the strategic/logistic interest of the industry to establish a production site at the specific site location, the product portfolio that the industrial partner can put into the business. The agreement reached with the company was such that only the FDG of the partner could be produced at the site. In general, having one sole product in the site portfolio can be risky, especially considering that the "attitude" of clinicians toward PET tracers may change over the time, and preference may be given to innovative products. On the other hand, the demand of FDG was pressing at the time.

A larger product portfolio would be a reasonable insurance of duration and profitability. However, GMP is more severe when multiple products are manufactured at a site, and this aspect was evaluated in depth, in particular in our situation, in which no experience could be used to face the challenge. Another risk is that other competitors may come up on the market and jeopardize your and partnering firm's ability to stay on the spot. On the other hand somebody could reply to this "welcome to the industrial world!"

Once a general convergence on the above was achieved, the following steps aimed at defining the project that would bring the old radiopharmacy to the new status of industrial production site.

The industrial partner was well trained on the analysis of third party's situation and had familiarity to get all key information on the site: it had to present a risk assessment to the corporate top management and try to demonstrate the project feasibility. The public partner usually benefits from this approach, quite far from a research project, and has the opportunity to learn about industrial evaluation landmarks and priorities.

The signature of a general No-Disclosure-Agreement (NDA) marks the point in which the team of the radiopharmacy comes into direct contact with the manufacturing process(es), while the industrial partner gets direct access to the site and full knowledge on layout, logistics, and personnel. The first visit was a first-hand on-site analysis, aimed at providing a quick go/no-go answer, mostly based on site location and building structure. Logistics of the site is critical with PET radiopharmaceuticals: facility access and connections to the road system may be a critical issue, e.g., when the risk is high that the short-lived product remains jammed in the city traffic. Following the exclusion of serious constraints that could stop the project, a more structured team from the industrial partner visited the site later on, with the task of preparing a list of immediate actions needed to adapt the layout to the industrial process and have the site passing a licensing inspection by the health authority. It is the moment in which the radiopharmacy team begins to face the true meaning of compliance with the rules applicable to the pharmaceutical market, and the different meaning of "product" between the previous situation as an in-house radiopharmacy and the future role of radiopharmaceutical manufacturing site. To try and better explain this thought, it may be useful to analyze the context in which my team and I were.

In Europe, compounding regulations differ from country to country, although an effort is underway by the European Commission for Pharmacopoeia to align standard references and have a uniform basis for quality control and quality assurance of products have clinical use. Almost anywhere in the EU, the in-house compounding of radiopharmaceuticals is based on national rules, and the legitimacy of the preparation is linked to a personal request issued by a medical doctor for a specific patient; i.e., a nominal prescription is issued for the patient(s) and the use of such a preparation remains under the direct responsibility of the prescribing doctor. The industrial manufacturing responds to different rules and regulations, namely the current Good Manufacturing Practice (GMP) defined by EU directives, which are incorporated within the legislation of each member states.

In Italy, the compounding of radiopharmaceuticals is regulated by the "Good Practice for Radiopharmaceutical Preparation in Nuclear Medicine," (GPRP) which was established as a National law (issued in 2005). This law covers both the preparation of licensed radiopharmaceuticals and the products required for a specific patient, under the prescribing physician's direct and personal responsibility. This kind of pharmaceutical preparations are performed at nuclear medicine departments, under the responsibility of a nuclear physician.

GPRP rules have a lighter architecture than current EU GMPs, which have a bulk of detailed chapters and annexes covering any of the specific aspects or tasks applied to the drug manufacturing process. Although many points of GPRP reflect or recall the cGMP approach, in general, the body of the text leaves some more freedom of implementation and adaptation to the local organization. In many cases, the law itself provides just general statements and indications that have been implemented into operational documents and protocols by a cooperative effort involving the competent national professional associations. GPRPs have deeply changed and promoted the quality of nuclear medicine departments throughout Italy. However, GMP and GPRP move on different tracks, in full respect of their final goal: industrial manufacturing and in-house production.

Beside the extended architecture of the pharmaceutical quality system and the impact on the site, the legal responsibility is more compelling in a GMP-operated site, because it is directly connected with the liability issues of an officially authorized drug. GMPs are enforced as a law, and severe sanctions are defined for all those who manufacture a drug (a radiopharmaceutical) by the use of unauthorized, uncontrolled, or biased processes.

The delivery of an imperfect product to a customer is equivalent to put a counterfeit medicine on the market. In fact, it is intuitive that placing a drug on the general market, means that it is available to any patient who qualifies for a clinical indication. A lower risk situation is involved in the preparation and release of a prescription which has been identified for the needs of a specific individual and will be used under direct control and responsibility of the doctor. Actually, beside the full medical control on the patient, in this case the general risk is mitigated by the smaller number of patients and the exceptionality of use.

Furthermore, the size of produced batches at an industrial (GMP) manufacturing site is often high, and the number of patients is greater and possibly distributed at different locations; therefore, the control over the process must be more accurate and more stringent rules must be applied to the supply chain. This is why GMPs are established and amended, whenever it is necessary to improve quality and safety, through regulations, directives, technical guidance documents, technical reference annexes, etc., and provide strict guidelines of what can or cannot be made, or must exist at a site where drugs are manufactured. This applies to any aspect of the process, from layout, to personnel, operations and controls, conservation and distribution (i.e., good distribution practice).

All this controlling effort (and sanctions) does not need to apply to GPRP, which is governed (and sanctioned in due cases) by different rules and regulations, possibly in line with compounding hospital pharmacies. GMP inspection and vigilance are performed by the national health authority according to official protocols aiming at ensuring an equal approach to product/site evaluation throughout EU, while local or regional reviewer committees (inspecting hospital pharmacies) should take care of GPRP control at radiopharmacies.

This needed to be well explained to the radiopharmacy team. Indeed, learning that the manufacturing process was so rigidly disciplined was a quite shocking experience for the team at the site and had the effect of spotting the light on the importance of a good start, in which any critical point was duly analyzed. An advanced site review was conducted, having in mind the manufacturing process described in the MAA (Fig. 6.1, From a local PET radiopharmacy to a GMP manufacturing site). It yielded a "gap analysis" of the current versus the final (GMP compliant) site situation. From this gap analysis, a cascade of key actions were identified, in particular those that may be a source of attrition in bringing the transfer of the manufacturing process into the candidate site to a success.

Feasibility evaluation of building adaptation came at the same time: for instance, the complexity of the layout modifications, the environmental air treatment (HVAC), and the installation of advanced equipment. The Pisa site had recently been built; however, it needed to be further adapted to be compatible with the new process and accommodate dedicated areas, like the warehouse and the packaging/shipping room, and accommodate a largely increased duty cycle.

For approximately 1 year, an intensive cooperative work was done by "waves" of industry GMP experts and managers at the site. Input questions/proposals had to find the way to receive practical and timely solutions and intermediate approval as well.

The public organization does not always have the technical skills on board. Thus, this phase turned out to be a quite hectic period, in which technical solutions needed to be outsourced and general authorizations obtained or planned by the site. In fact, a public institution may have a complex and distributed organization, and I was indeed in such a case. The gap analysis ended into the elaboration of a to-do-list,

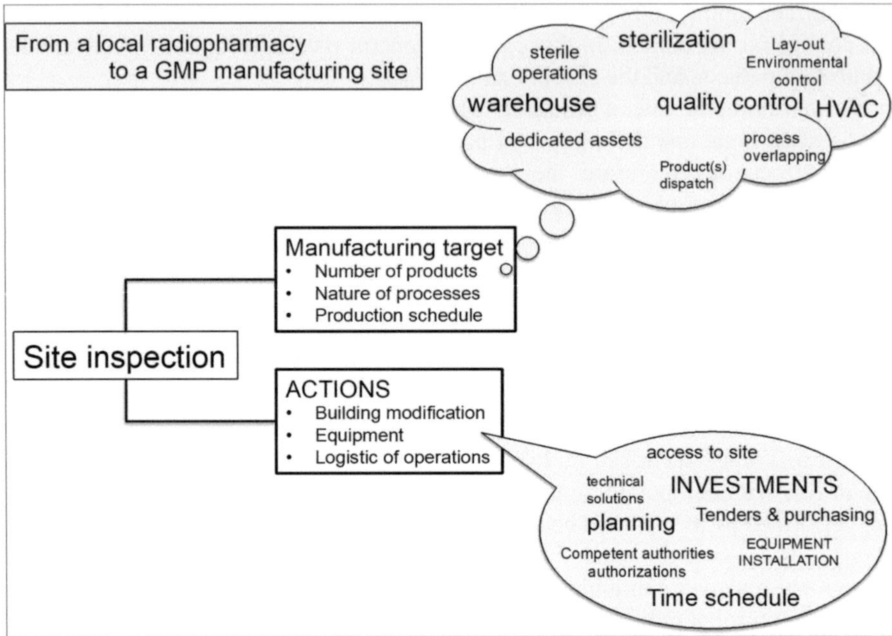

Fig. 6.1   From a local PET radiopharmacy to a GMP manufacturing site

which included a risk evaluation, to rank each item in the list as more or less critical, and a reasonable estimate of associated costs. In other words, the completion of this phase is a relevant preparatory activity before the full contract starts to be finalized.

A Personal Story: The Site Visits

I had a long experience of site visits, from visiting professors and scholars to official entities that could even close out the laboratory: the competent authority for radiation protection, the local sanitary office, the office for workplace safety, etc. Nevertheless the site visits by the industrial partner were always a bit more worrying. It was like being halfway between an evaluation at school and a medical examination. The first because, somehow, you get scores on a lot of items on which you still feel quite uncomfortable; the second because you would like to know what is in the mind of your interlocutors before they write their notes.

The analysis of milestones set out by the contract giver rises the anxiety level even more: for instance, layout approval, GMP readiness analysis, corporate QA review are always thrilling experiences. Moreover, the larger is the company you have as a partner, the larger will be the number of persons participating in these meeting: sometimes a few guys against an army! All issues are analyzed and described in a report, and soon I had to change my mind on the meaning of "obvious": it was obvious only what was written and documented, as it is stated by GMPs.

My team and I learnt what is the meaning of punctual inspection, and reporting as well! We learnt very quickly that when you tell your reviewer: "It will be fixed soon," the question "When?" would follow at once.

The site visits are the best opportunity for a live test on your ability to manage the site, with its newborn quality system, and face some new approach to life: first of all, risk assessment.

Risk evaluation exists also in the academic world, but rarely needs to be estimated on a purely predictive basis. The normal way is that data or measurements are taken, results analyzed by using suitable statistical methods of analysis, and numerical conclusions are drawn. How can the risks associated with a process or a site not yet in operation be estimated and produce information that tells whether the decision/action taken improves or not the outcome?

In the industrial real life, if risks exist, then failures may also happen. On top practical negative outcomes (a batch rejection, a failing control, a failing production, …), there is always a deviation from what should have been the perfect functioning of the system.

Then, the root cause has always to be identified if the risk has to be eliminated or at least mitigated. Why has the occurrence happened? Here it was a surprise: it would be misleading stopping just at the first answer that comes to the mind, for instance: "it was an operator's mistake." In fact, this may still be consequent to other reasons, and the root cause (the real cause) has to be found. The "5 WHYs" method comes to aid. It is an iterative interrogative approach, in which who analyzes the deviation asks "why?" to the occurrence and repeats again the question to the following four answers:

" The control failed" … "WHY"?
   "it was an operator's mistake" … "WHY"?
   "He did not respect the SOP" … "WHY"?
   "The training was insufficient" … "WHY"?
   "The topic was not clearly covered in the SOP" … "WHY"?
   "Section 3 of the SOP needs to be rewritten to avoid further recurrence of the deviation"

Once again, this was completely far from my mindset as a scientist, and it took long to correctly implement this approach in the normal handling of deviations: it sounded so crazy and annoying, in the beginning!

This example tells you much also on the different approach between research and industrial production. In the first, the process has to be developed and improved; therefore, the variables that have an influence and weaken it are to be identified. In the second, the process has been validated and nothing from outside should interfere with the expected optimal result.

The Writing of the Contract

When academy and industry approach each other to discuss about "products," a different cultural experience, goals, and even a glossary mismatch can complicate effective interactions. Therefore, acquisition of education on GMP procedures and industrial "way of thinking" may be a valuable benefit for the public partner because of the special spot-light existing on the product, its history, characteristics, and risk of defects. This is particularly true if there is an interest in creating a bridge between basic research and drug development, in fact the drug world is a very complex space, in which ethical and economical aspects converge.

When the parties start the discussion and enter the contractual phase, no political, technical, or logistic critical issues are anymore present; obviously, a delicate and complex phase of negotiation starts between the two partners. Any contract covers the direct interests of the parties, therefore each single partnership can reach a different level of agreement and cover peculiar aspects in response to actual goals and motivations. However, some general principles remain valid and can describe the underlying dynamics that may operate.

In general, the contract will compose expectations, business strategy, economical considerations, and risk assessment as well. For instance, the motivations of the public radiopharmacy may go beyond the starting of a profitable business. Achieving a full-GMP compliant facility—mentored by the industrial expert advice—may be an added value, even in the case that the partnership with the industry might stop at a certain point (e.g., when the contract expires or because the industrial partner withdraws, due to changes in its business strategy). The public body may even be interested into promoting the transition of a high-tech/high-cost internal department, such as the radiopharmacy (cyclotron, hot-labs, equipment, personnel), toward an economically independent organization, that may operate as a private firm, and still be under its control.

The industrial partner may see the low duty cycle at an in-house production site as an opportunity and a valuable prospective resource. A radiopharmacy, working for in-house production, delivers its FDG batch usually once in a day, around 7–8 a.m. The activity delivered can accommodate one full PET/CT diagnostic service, even when two scanners are working on-site in parallel. Normally, the PET diagnostic service closes in the early afternoon, as a balancing between nurse/radiographer shifts, the time allocated by nuclear physicians to evaluate the exams and prepare the reports, the stress for patients, who may have difficulties in keeping the fasting state needed by the procedure at later hours. This means that, for instance, the radiopharmacy personnel start activity between 5 and 6 a.m., work till about 8 a.m. on FDG production, and then assist the radiographers in patient dosage preparation, or even work on a second run, in case small batches are produced. Bringing the site from one production per day to two, three, or even four extra runs, for the industrial partner means having the chance to exploit the unused working time of an existing facility, and the costs to make it GMP compliant can be easily balanced by the marginal costs of the plant and the full duty cycle operation.

The industrial partner may also be interested into additional targets, not necessarily limited to costs saving and improvements in the logistic of production and distribution. These may range from having access to possible innovative product that are under development at the site (e.g., new products in the pipeline, receiving research support in clinical trials), to exploiting future site development such as benefitting from the background of scientific and technological expertise existing at the site and nearby (e.g., as it is common at a university, a research campus, a technology park) to test new manufacturing processes and products.

Therefore, the contract may include additional items, whose value is bound to each party's strategic vision, beside the simple definition of who pays what. Anyway, the balance of payments will cover any specific situation or settlement. The contract itself may deal with many different aspects, and a convenient approach to keep it clear and well specified is to have a main body (the general framework of the agreement) and a number of annexes, each covering specific topics in a practical and detailed way (Fig. 6.2 contract definition).

A public body, such as a nationwide organization, can operate through a cascade of institutional levels. A central administrative headquarter usually deals with the general policy and keeps relationships with the national government or its reference structure (e.g., a ministry). Then, it may operate through directorates, department, institutes (clinics), which may be diversely distributed over the country. In principle, each of them may have a different level of independency and decisional autonomy. The person (the director) of a public institution who takes the legal responsibility of the contract, often operates as a delegate of the legal representative (usually at the headquarter). It is rather intuitive that when a contract is proposed to a peripheral

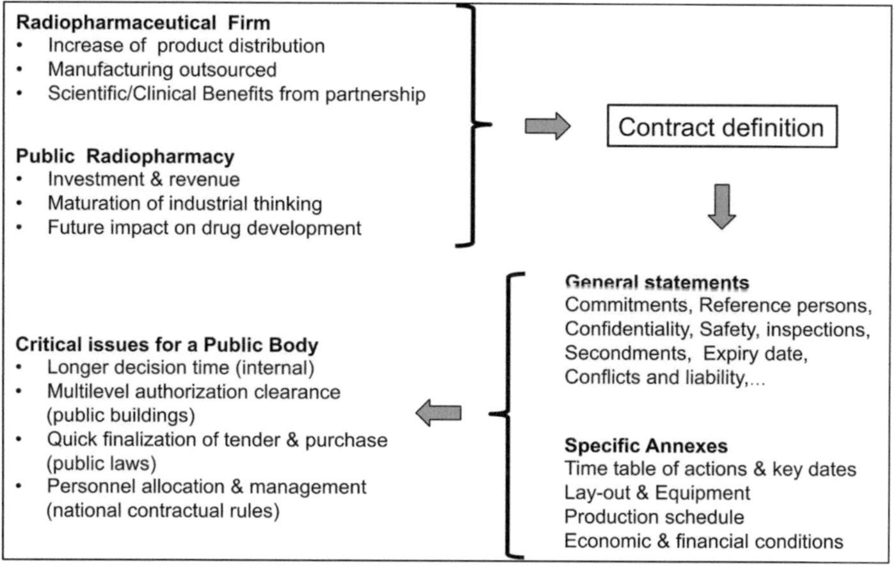

Fig. 6.2 Contract definition

officer, it will involve higher levels for judgment and require approval by the relevant central office(s) (e.g., legal, purchasing, accounting, …) before it receives a green light.

The review of the contract by a public body is complex and an analysis shall consider, beside the economical aspects, whether any conflict exists with both the general policies of the public institution, and the regulations that are applicable to the public administration. A more or less high burden of bureaucracy has to be experienced before the formal frame of the contract is set out. This refining activity may last few months and will lead to a fully agreed contract with all appended annexes approved.

In the end, the general part of the contract includes the description of the agreement (mentioning the relevant annexes for detailed descriptions and application of rules), the key persons nominated by the parties, the contract duration and the conditions of resolution, the management of disputes and controversies. It may also include a point concerning penalties, to be applied in case the parties would not respect contractual obligations and targets. This may be a most critical point for a public institution, that is much more acquainted with a cooperation or collaborations than with the delivery of a service.

The clear and exhaustive definition of the annexes, which are more related to operational details, are discussed more locally between the parties; at the site, they are largely demanded to the key persons, in particular the director and the site managers, and the central offices of the public institution will only "legally debug" the contents.

In the end, a set of annexes covers the reciprocal commitments, from the general schedule and timetable, to the expected site performances, and organization (materials, personnel, and equipment management). The list of possible annexes and their detailed contents may vary according to the specific case; therefore, the examples that follow describe my personal experience.

Annex on Layout Schedule and Timetabled Actions

The key points of this schedule are hinged on:

- Approval of key documents (e.g., layout design, technical drawing, project clearance by involved authorities)
- Execution and fulfillment of actions in the building (layout modifications and integrations)
- Equipment purchase and installation
- Site preparedness for operation

An industry can decide over a short time to modify a building (or even raising a new one) provided it is on its own property: briefly, "time is money." The situation may be different in a public body, in which a more complex hierarchical organization exists, and the enforced rules privilege the "ex ante" versus "ex post" evaluation of responsibility.

The director is not the owner but the "tenant" of the infrastructure, and any change or modification to the actual situation needs to go through more or less complicate procedures, which may involve one or more levels of authorization. Of course, this is a country-wise problem, but in general it will depend upon the overall organization or, saying it differently, a real-world situation should be kept in mind by the parties when they put together the timetable that is appended to the contract.

In the EU, a general European tender law—implemented nationally, and thus with modifications from country to country—has been enforced: this poses a number of rules, constraints, and limitations, which are justified by the need to ensure the maximum latitude on tender participation by firms. Moreover, an equal structure of offered goods or services must be adopted, so that the comparison of different proposals is facilitated, and the public body keeps the best chance to achieve the most rationale expenditure. This is remarkably fair from the view of commercial ethics, and it is also politically correct, but often it translates in heavy and time-consuming procedures before the purchase is finalized. Furthermore, the plethora of documents, to be presented before and after the order is placed, may also become a reason of complains and further delay. Therefore, such a "side condition" has to be duly regarded in the annex when the public body takes responsibility of tenders and purchases. In fact, such a decision may be highly impacting on layout modifications that have to be made to the building, which is unfortunately a frequent situation even in a recently constructed building.

The compliance with GMP relies on strict guidelines prescribing the characteristics that working spaces and the connected air treatment system (HVAC) should (must) have. The latter in particular includes a precise direction of air flows and a cascade of pressures and ventilation rates across rooms, which varies in accordance with the planned activities. The air conditioning system of an in-house radiopharmacy could be insufficient to the new purpose. In a similar way, the radiopharmacy layout usually needs revision; so that the areas dedicated to warehouse (sometime not present), production or quality control, are separated, and the transit of personnel and materials is well designed, without crossing of incoming and outgoing flows.

The increase of the production capacity, or the different pharmaceutical approach to the product, may even require that the core equipment of the production, i.e., the hot cells, are modified or even replaced to accommodate the new process. This was the case when a process, which was based on the autoclaving of FDG in its final primary container, was replaced by a new one that included an open vial sterile filling. The adoption of a validated process often entails specific equipment and instrumentation, e.g., those qualified during the manufacturing process development, which have to replace or integrate those already existing at the site. Once again, the timing of purchase has a relevance, but even the a priori selection of a supplier (e.g., the one qualified by the MAA holder in its supply chain) cannot be easy. Due to the European regulation discussed above, the public partner must contact and evaluate a multiplicity of suppliers (not less than five!) and make a price/quality evaluation that become public in its conclusions. Direct purchase from a single supplier can then be finalized only if justified by adequate motivations.

Sometime the industrial partner can take care of the more complex actions, such as building revamping and purchase/installation of large equipment. In this case, the economic balancing may be done in the financial chapter of the contract, provided the industrial partner has enough margin to follow this way.

Additional use of the equipment and the site by the public Institution can also be arranged in the contractual conditions. This annex can also include some strategic views. For instance, the layout can be dimensioned, within the limit of the available budget, aiming at the highest production performances of the site, so as to have room for further developments, without new actions or revamping on the infrastructure (additional equipment can come later, when they are necessary).

Annex on Production Startup and Validation Schedule

This annex covers the description of the steps needed to reach the GMP status of the site. It is the second half of a single match, with a direct connection on what is the outcome of the first half. In fact, this schedule starts when the layout and equipment (at least the core instrumentations) are available and ready to come into operation, after successful qualification.

If the general contract includes the manufacturing of more than one product, a scaled solution, i.e., one product at a time, can be adopted to simplify the track toward regulatory clearance. Even if such a choice will make longer the time to achieve the full site operation, it is wise to separate the inspections so that the team and the inspectors can focus on a single process at a time. Of course, the independence that can be assured between the different production lines, and the resources of personnel that can be employed can also influence this decision.

As a core activity of this schedule, the owner of the Marketing Authorization (MAA) formally transfers the licensing Dossier(s) to the team at the site. In this way, the key functions at the site (the QP and the managers in charge for QC, QA, and production) will have access to all chemistry, manufacturing and controls (technology transfer team) documents that will be implemented and adopted. By that time, the site must have enforced a QMS that ensures data recording and control according to GMP. It is a bulk of documentation and SOPs that have to be prepared by the site and reviewed by the industrial partner (as MAA holder or MAA licensee), to secure coherence with the relevant industrial process. More in depth, it is a different way of working with respect to the previous radiopharmaceutical practice. It is a considerable effort, therefore the steps described in this schedule should take care of this and drive the QA team at the site through the activity moving from the paperwork to the practical operation.

The "rules of the game" for such a cooperative activity, which involves the team as a whole GMP squad, is comprised in an independent agreement signed by the parties, the Quality Assurance Agreement (QAA). This is usually mentioned into the contract, but it is finalized later on, when a more accurate schedule is possible, and the site is starting to function. The QAA clearly defines two major points: the relationships between the partners' QMS, and the calendar and contents of the validation plan.

The schedule on validation provides a calendar of the actions to be taken before the industrial process is transferred to the site. A Validation Master Plan (VMP) guides the site through a number of defined steps: installation qualification, operation qualification, and performance qualification of equipment. It includes aspects that are often new to the radiopharmacy team, such as the starting and management of a warehouse in which a qualified supply chain defines material acceptance and storage.

Microbiology is a vast territory dealing with environmental monitoring and control. The smaller number of daily preparations and the different regulation can make this type of control less demanding in compounding radiopharmacies than GMP sites. In industrial manufacturing, this is a very delicate point, in particular when sterile filling is included in the process, which is carefully evaluated by the inspectors. The Health Authority (HA) takes the risk of microbial contamination of the product (especially if injectable) very seriously; such an attention ranges from the scrutiny of SOPs, to gowning (special coverall and personal sanitation are adopted), and control and sanitation of surfaces and equipment, in particular when plate reading after exposure is performed on site.

The VMP is a dense and composite plan, pursuing the full process validation and its goal is the production of three lots manufactured according to the licensing Dossier. The manufacturing of three lots is the milestone that needs to be reached before the application can be presented to the HA for the site inspection and GMP certification.

As a matter of fact, the definition of the timing in this schedule has a parallel effect on the design of the overall training at the site, and in particular the training of the operators that will be assigned to the production and quality control departments.

Annex on Batches and Site Operation

Obviously, the definition of the number and size of batches per day has a prominent importance in the contract, as they represent the main content that is related to the business profitability for the parties. However, this schedule is usually easy to write, because the general targets of production and site performances have been outlined during the preliminary negotiations and have consolidated into the contractual agreement later on. The industrial partner has already made an upfront market analysis and put on the table of the negotiations its selling targets, and a possible plan of expected service evolution in the future. The site had time to express its view and take its commitments. In general, the production (frequency of runs, manufacturing days, shipment time, batch size) grows over a defined period, so that the increased pressure on the site can be balanced by the improved experience and confidence with the process. The more delicate issue in this schedule is to set up the logistics and the organization of the product dispatch and delivery.

By this time, the logistics associated with the delivery of the product has to be either implemented or at least defined in its operational lines (terms, limits of service, and responsibility); in particular because it has a connection with a new actor, unknown to the site: the customer. This connection moves along a red line that link the order with the delivery. Once again, this is a completely unexplored territory for the team at the site.

It is common practice that the owner of the MAA or its licensee, i.e., the industrial partner that has the function of contract giver with the site, keeps the contacts with the end users (physicians) and customers (purchasing officers).

The purchasing of PET diagnostics by public institutions, that is the majority of hospitals and clinical centers in Italy, has lengthy and complex procedures and the site in general has neither interest, nor a valid organization, to deal with it. The contacts with the customers are almost always kept by the firm's commercial and tender and quote services. The different role played by the physician (care of patients) and the hospital purchasing office (handling of administration activities) may add additional complication because their instances may move at different speeds.

The handling of orders is a key point in the overall organization: many clinical sites wait the very last minute to close their booking service; furthermore, changes can even be communicated even later, following emergency requests by physicians. The orders received from the different customers are then gathered at the industrial partner's office and grouped according to logistic considerations, such the geographical distribution of destinations, the road conditions, the requested delivery time. The grouped orders are then converted into one or more production schedules that are transferred to the site. This may end up in a short notice to the site, usually the afternoon of the day before the production is due, i.e., few hours before the operations start. On the other hand, the nature of the product does not allow any stock, and the whole organization must be able to cope with this unavoidably stringent flow of orders.

The preparation of the production schedule has not yet completed the task of the logistics organization. The shipment of radiopharmaceuticals to their final destination is complicated by the mandatory application of the rules for the transport of radioactive materials and goods (ADR), which impose the use of certified vehicles and special labeling of the products (beside those needed to identify products and consignees).

By the time the production starts, the fleet of vehicles of the courier service has to be decided, organized, and booked. The carriers must be at the manufacturing site in due time and take care of the timely delivery of the product to the different destinations as soon as it is available.

Who-does-what is here a relevant point to be precisely defined in the Annex, even if the logistics is demanded to the industrial partner. The contract giver has experience in goods circulation, while the public partner will hardly be able to have a dedicated staff to manage this activity. Efforts by the team are focused on production, which includes packaging and labeling of the shipment containers. This is not a trivial activity because the labeled primary container (a vial) lays into a labeled secondary one (lead or tungsten closed pot), which in turn has to be placed inside an anti-shock plastic case (Type A, certified ADR container), which is finally sealed and labeled. One single lot may have multiple shipment destinations, i.e., multiple packages to be carefully handled and avoid mismatch.

The Annex on production includes also the management of critical situations that may occur at the site, such as production failures or batches with a yield lower than expected (insufficient to fulfill the clients' demand). Then, a protocol describing the communication and cooperation tasks between the parties is included in this Annex. Future agreements (e.g., detailed in specific SOPs) will duly describe how to deal with all emergency situations, including courier no-show, delay, accidents, traffic jams.

Annex on Personnel

The production schedule has a deep impact on personnel, beside the company's strategy and marketing needs and the site expectations in monetary terms; personnel organization is therefore an important and complex aspect in the contract, in particular on the side of the Public body.

It is quite natural that the staff already aboard the radiopharmacy joins the manufacturing team of the future industrial site. However, the increase in the production targets, and the extra tasks of manufacturing introduced by adopting GMP operations, generally require a larger team. There is some reluctance in having sites in which there is a mix of personnel coming from different employers (e.g., the industrial partner and the public body). Working conditions (e.g., position, salary, working hours organization, and extra time) are regulated in many countries by national agreements, that can be legally binding and introduce differences across categories, sectors, and eventually public vs. private workers. Therefore, the economic conditions set out in the contract usually accommodates specific funds to allow the public partner to hire extra personnel, who will be employed according to the contractual rules applicable to the public administration.

The short-lived nature of the PET radionuclides indirectly plays a role on the selection of personnel. In fact, the production must be compatible with the timely utilization of the product at clinical centers, that usually start their activity with patients early in the morning. Therefore, night hours are included in the normal span of time for manufacturing, in particular when more than one production runs have to be scheduled, or the end users locate far from the manufacturing site, so that an early shipment is needed.

A normal start of the operation at a PET tracer manufacturing center is around 9 p.m. of the day before the dispatch of the product. This timing is calibrated on fluorine-18 labeled drugs (such as FDG, fluorocholine, and fluoro-DOPA), which have been in my past experience. In fact, a three half-life time span (about 5–6 h) is usually considered as the practical maximum time window to finalize the delivery to the farthest customer. To ensure a delivery between 7 and 8 a.m., the product has to leave the site around 2 or 3 a.m. Therefore, production has to start a few hours in advance, to allow for cyclotron operation (1–2 h target bombardment), production setup, and QC equipment starting and calibration.

Making public personnel to accept a night shift may not be easy, and may not even be contemplated by the aforementioned national agreements. The national agreement for employee of my institution did have an article on shifts but, in general, working on a night shift is not easily accepted by a public employee. Therefore, a preliminary activity, done by the director and myself as site manager, was aimed at exploring and discussing the organization of the shifts and the associated benefits (monetization, extra rest time, work-time reduction) with the personnel. Of course, this had the limits posed by the applicable laws: for instance, providing extra money outside the economic conditions contemplated by the national agreement was unfeasible for public employees.

Indeed, it can be a tough task for a manager to keep the personnel motivated and in line with the due operation schedule. Nevertheless, there are people who like working by night and have extra rest time when everyone is at work or simply more time for children care, although in my experience this does not last forever. A possible solution is to have a team larger than the strict number of operators. This will reduce the stress on personnel and on the site manager as well, who can more easily accommodate resources during holiday periods, illness, or other unpredicted mishappening. A wise person might say this is a worsening of the economical balance of the site. The answer to this point is complex. Many public bodies, including mine, have a centralized chapter that covers the salaries of permanent staff employees, so that they are not directly charged onto the budget of the institute/clinics to whom the site belongs (like instead other direct costs and consumables). The rationale behind this approach is that the mission of the permanent staff of a public research body is dedicated to promote basic and applied research and support actions leading to public benefit, technical innovation, and social promotion.

Thus, a part of the salaries are not charged on the budget of the future manufacturing site and this increases the margin that can be used, within the available budget, for hiring personnel with temporary positions. Therefore, the additional personnel for the site can be hired and charged on the costs of the contract and included in the negotiation with the firm. The industrial partner may also be happy to avoid the need to hire and use its own staff at an external site and be willing to support this solution for the best outcome of the project. So, it may be a win-win situation, provided you find and keep the persons on track!

In fact, one more complication of working at a PET manufacturing site is that it can be considered a strenuous work. Beside the obligations due to the conduct of operations in clean environments (e.g., special and additional protections, complex gowning, multiple garments changes, sanitation), radioactive compounds are involved into operations. Even if the field of radiation protection applied to PET radionuclides has reached high safety levels, including the development of shielded and safe high-tech equipment, additional procedures and controls are to be enforced.

This makes the life of personnel a bit harder and often may be complicated by the post-Chernobyl syndrome on radioactivity exposure fright. Monetary compensation for difficult working conditions is available in some countries (Italy is one of them) and can help to keep workers motivated. Finally, the operators have to move radio-active samples and products in shielded containers, weighing up to about 20 kg, which have to be lifted and handled repeatedly during each single production batch. Therefore, solutions have to be foreseen to avoid accidents due to the falling of heavy objects and fatigue in less exercised operators. Not all of the issues discussed above necessarily are mentioned in the text of the Annex on personnel; nevertheless, it would be wise to have an early attention to these problems, possibly with special care in the design of the layout so to avoid hurdles, and facilitate the job to the operators with the use of pass-through boxes, belt conveyors, and lifting equipment.

6.3   From Words to Deeds

The Contract Implementation: Part I, Preparing the Ground

After the cheers and toasts following the contract signature by the parties, a hectic period starts in which the empty boxes in the timetable have to be filled. The first step is fixing the organization at the site, and this may require some time.

The director (legal responsible at the site) must appoint the Qualified Person (QP). This position is not necessary in a radiopharmacy, but it is mandatory to estab-lish a GMP site. The QP will take full responsibility of the compliance of both the process and the product with the GMP, and the approved specifications, as dictated in the MAA. In many European countries, the QP must be an employee of the legal entity managing the manufacturing site and cannot be a consultant. Therefore, he has a prominent role in the operation of the site and does not necessarily coincide with the site manager. The director must ensure that the QP is hired, unless an employee can qualify, having the required curriculum and experience. In Europe, the title is issued by the HA, after having assessed that the candidate has a valid cur-riculum, which includes a specific academic title (master university degree in phar-macy, chemistry, biology, or medicine) and an official working experience at a licensed site: from 1 to 2 years on-the-job, according to the academic degree. Therefore, the time gap to have a QP at the site must be considered in the general schedule of site construction and approval.

The names of the key persons, and future replacements as well, need to be for-mally communicated to the industrial partner and will be the official GMP organiza-tion chart at the Site. Usually, the Site Manager identifies the three key persons having the responsibility of production, quality control and quality assurance, while the director makes the formal appointment. These positions may have been already present in the radiopharmacy, nevertheless the persons in charge need to be fully aware of the GMP meaning of their role, and an accurate job description is prepared for each position and signed by the nominees. In Italy, the personnel assigned to production or quality control departments of a GMP site, and not only the managers, must independently operate from each other to ensure full control of operations.

The personnel of the radiopharmacy, be it grown in a hospital or in a research center, must be introduced to the GMP, so training and education are extremely important to ensure full control of the many future actions that will be addressed while implementing the contract. The good radiopharmaceutical practice, that have been enforced at the site, may not be sufficient and even be a possible source of mistakes due to mismatches between the two quality systems (GPRP vs. GMP). It is a wise solution to have the firm's trainers doing this job. In fact, the GMP existing at the site and that adopted by the firm will need to communicate efficiently and effectively when the contract is operational.

The adoption of GMP is not immediate: there is a different mental attitude between the personnel working in a compounding radiopharmacy and that of an industrial manufacturing site, in particular when changes are deemed necessary. This may even be worst in research centers; in the field of research, protocol changes are decided or introduced as a part of the normal project evolution. The protocol is amended, and the new procedure simply replaces the older one. The entire process remains dynamic. In GMP, the process is already defined and approved, and the project for its modification must be established beforehand. Any further change must be justified and approved again. Quite a reversal in the habitual process!

As mentioned earlier, there is a variable situation on how good practice is applied to compounding across Europe. It ranges from industrial full GMP, including inspection of sites by the HA, to GPRP and very light inspections (if any). Therefore, the adoption of GMP by a site may be cumbersome and quite often many SOPs, if not all, are imported from the quality system of the industrial partner and adopted at the site after the modifications that are necessary to adapt them to the local situation. This will speed and facilitate the building of the Quality Assurance Agreement (QAA) that needs to be signed by the parties.

The intense paperwork on SOPs moves in parallel to the works that have started at the site; then, coping with the schedule is the challenge. The risk of delays is hanging on tenders and purchases, but also on the delivery of materials and instrumentations. So, it is a tough task of the site manager and the director to ensure a careful follow-up and provide prompt reaction to any critical situation. The labs and the offices near the site keep their normal operation, and the risk of interference is high; therefore, this is probably the most critical phase of the project and a continuous and precise interplay must be maintained between the team, the general contractors, and the engineers in charge for instrument rigging and installation. A very complex orchestra that the team at the site has to direct at best.

In parallel to the activity going on at the site to prepare the infrastructure and install the equipment, the team is concentrated on starting the QMS. The compilation of procedures (SOPs) and the definition of the core documentation of the future process take time and a lot of work. The quality system has general rules and not a frozen format: it has to be adapted and applied to the future specific process and layout. Such general principles shall be respected during the drafting of working documents, which must be written in a balanced way, not too restrictive nor generic. The process and how this is implemented at the site (i.e., the planned technical and operational choices) need to be analyzed and broken down into components that have to move together in a synergistic and coordinated way.

The functional destination of any single room of the site, the flow of materials and personnel, and the working ranges of pressure provided by the HVAC are defined with the layout design, attached to the contract. The full list of instruments and special equipment needed in the process, with their relevant specifications, has been set up in the discussion between the Parties.

The team moves along a learning curve that includes cultural milestones that go beyond those in the time schedule. For instance, a relevant point to the team is learning the concept of equipment qualification. A similar activity is usually required at a radiopharmacy when an instrument should be included in the daily operation. The equipment is often controlled for completeness, its functions are tested, and the performance checked against the manufacturer's demonstration criteria or the standards adopted in the literature. This process in considered very critical in GMP, and the qualification of an instrument requires a very precise and detailed track that always includes installation qualification, operational qualification, performance qualification and incorporates the data integrity demonstration for electronic and computer equipment. The GMP qualification process leads to the demonstration, in separate and successive steps, that the instrument is able to be utilized "for the intended use." Each step has a specified set of tasks to be carried out, and a dedicated documentation to be prepared and filed.

The GMP tests of good performance demonstration are finalized to the control of the analyte (that is the one specified product or reagent of the process) and not only the demonstration of a correctly performing analytical instrument. The same philosophy applies to all equipment and any critical device at the site.

The radiopharmaceutical process drives the complexity of the site layout and its operation. In fact, two points should be kept in mind: first, PET radiopharmaceuticals are injections, therefore the final product must be sterile; second, more than one product vial should be normally prepared in the process, i.e., the bulk product must be fractionated and dispensed. Sterility has to be ensured; the chemical nature and the thermal stability of the formulation decide whether this can be achieved by autoclaving or sterile filtration. Over the time, I had experience with both approaches.

The autoclaving of the product in the final, closed containers is performed after that it has been formulated, fractionated, and dispensed into closed vials, each loaded with the desired amount of radioactivity, which is defined in the production plan. The sterile filling is performed in a sterile environment in which the open vial is loaded with the radioactive dose as defined by the production plan, and the vial is sealed before leaving the sterile environment. For PET radiopharmaceuticals, this two alternative routes translate into very different solutions, for both the site layout and the equipment to be installed, and there is a real change in the overall complexity. In fact, the first process can reasonably be conducted in a Class C environment, which is easier to maintain and monitor. The sterile filling requires a Class A environment for the filling station, and a Class B one for the entry of materials and consumables. This working condition is usually achieved with the aid of sealed isolators that can ensure the necessary aseptic working spaces inside a Class C background.

The sterile dispensing has a more general application. It works with all products, even with those that are temperature sensitive and is compatible with a wider type of formulations, including viscous fluids. On the other hand, it requires much more work on the side of the number and complexity of operations (environmental controls, instrumentation control and maintenance; media fill; sanitation, operation and reset, SOPs) and hence of personnel training and learning time. The qualification of the clean room and the sterile process (isolator or sterilization equipment) are critical key points, and in general heavily occupy the last step of the site qualification. When this milestone of site qualification is reached, a major milestone has been achieved; the process can be transferred to the site. Now, the completion of the production of the three validation lots is the new goal.

The Contract Implementation: Part II, Aiming at the Inspection

There is always a learning curve to follow; not only to understand the spirit of having the process under control, but, more generally, the philosophy of quality applied to the pharmaceutical world. There is a sort of liturgy in the daily recordings and form compilation, once the site has started. This includes that the team has to get used to good documentation practice and record keeping. It has to be well explained and motivated to the personnel, to avoid that it is regarded as a deeply boring list of tests and controls, increasing weekly, until the site has completed the qualification and microbiological controls enter their routine.

The better the planning the smoother will be the reach of the target. During the phase of layout infrastructure completion, the team of the future GMP site has worked hard on the development of an internal QMS and the definition of a validation plan.

Any single part of the process (and its transfer on to the site) is analyzed on the basis of a risk assessment and with the assistance of firm's corporate QA and technology transfer experts. However, keeping many points of contact between the QMS of the site and that of the firm is a very wise decision: from raw material (specifications, controls, approval, and storage) to finished product specifications (analytical methods and acceptance criteria), to environmental controls and museum samples, the way of handling and management have common features between the parties. The same should apply to the management of deviations and change control, in consideration of the strict liaisons with the specifications of the product and the QAA signed with the MAA holder.

The QAA and its possible technical annexes follow the contract signature. It is a technical document, centered on the specific manufacturing process of a product, that describes and documents the exchanges of information and the reciprocal roles between the parties. It should be signed and put into operation before the start of the practical activity at the site. The QAA covers a wide range of topics; all of them are functional to have full understanding between the parties on all issues concerning the manufacturing process, in particular the level of interaction and communications between the parties, for the product(s) and services that are reciprocally supplied. In fact, the site will manufacture a product licensed to the contract giver, the contract giver can supply raw materials, reagents, or services to the site. For each specific action (and possible related tasks, controls, and specifications) that is specified in the QAA, it is defined and agreed which party has the responsibility, whether a preventive communication to the other partner has to be made, or who has to provide timely information on specific issues.

A common glossary provides shared definitions and mutually recognized meanings to sensitive issues of the process. The QAA can even specify particular agreements, limitations, and conditions that the parties decide to regulate in a formal way. This agreement can have effects on SOPs and manufacturing steps, such as supply chain selection and approval, analytical controls, calibration, finished product tests and controls, but also the management of situations that may impact on the whole product quality, such as handling of changes and deviations.

One of the most complicated sections to put into the operation is the warehouse. Usually, the industrial GMP rules on these points are much more severe than those adopted at conventional radiopharmacies based on GPRP rules. In particular, they differ in the qualification of suppliers, and in general the management of raw materials and equipment used for storage and control. Ensuring a controlled supply chain is a new task for the site, especially when the suppliers need to be inspected and evaluated from a QA point of view. Once again the cooperation coming from the industrial partner is a valid support to speed up the process.

The availability of approved materials at the site requires that the extensive set of procedures has been finalized and the flow of materials (and personnel) defined and in place. The establishment of the correct flows of materials, in and out of the warehouse, will require more than one training session on the record keeping, organization of quarantined and approved materials, and of retained samples. The QAA may help in this activity as it can provide the "justification" for using approved materials supplied by the firm. It is a good way to speed up the qualification of equipment and approve the analytical procedures, while the site suppliers are identified and qualified. Sometimes, specific materials can be temporarily or permanently supplied by the contract giver to accelerate or simplify the progression of the operations.

Communications between the team at the site and firm's QA and technology transfer team are intense in this phase, and reactions (corrective and preventive actions) to "deviations" must be discussed and corrected having a common attitude between the site and the firm. Quality assurance teams on both sides have much to work on this specific aspect.

Once the site and the equipment have been qualified, and warehouses put into operation preliminary productions tests are performed. The team quickly masters the chemistry/analytical operation of the process, as the chemistry of the process is often within the preexisting core competences, due to the previous experience. A hard period may be experienced instead in closing the training on sterile operations.

The operators have matured their experience with compounding, in which a closed vial operation is usually conducted and small batches (often one single vial) are produced. The new process requires multiple vials to be produced, with a more complex list of materials and operations entering and leaving the clean room and the isolator. All this has to be performed while keeping the sterile process in perfect conditions, both ensuring that the clean environment operations are smoothly performed and producing the required documented evidences per single batch. A friendly and efficient relationship with the Batch Record (BR) is the goal to be reached in this phase; in fact, a correct management of the BR testifies the essential level of skill the team has achieved with the process. Generally, the BRs are communicated between the parties to prove readiness before Technology Transfer (TT) is performed, as the final training step.

Technology Transfer (TT) is a full exposure of the team and of the equipment to the pharmaceutical process that will be conducted at the site. The technology transfer team of the industrial partner leads the TT; it has also the function of doing a verification of the team readiness and site suitability, thru an on-the-job evaluation of the GMP compliance achieved at the site. When TT is concluded, a formal documentation is signed between the parties that proves that the process has been successfully transferred on to the site.

The manufacturing of three lots, complying with GMP and product specifications and local laws, is a major site goal. It marks the time to start the interaction with the health authority and get the site licensing.

The application to the HA for the GMP certification comprises a set of documents dealing with the site QMS and a full documentation of the manufacturing process and has to be formally submitted by the legal representative of the site (the director). The HA inspectors will visit the site later, on the basis of the product specifications described in a registration dossier, and will peruse the site quality documentation, the VMP and check the completeness, robustness, and performance of the QMS. This is the most thrilling moment for the site team, in particular the QP and the managers, who will have a head-to-head discussion with the inspectors over multiple days.

The granting of the GMP certification to the site is a turning point in the relationship with the industrial partner. First, the formal track can be started with the HA so that the site can be recognized as a third-party manufacturer for the relevant product; second, the site can start operations for including additional products, if any, to its portfolio and get prepared to routine production when it will be possible.

A Personal Story: Striding Toward Site Operation

There is a basic difference between the practical view of the future by an industry manager and a scientist from academia. The latter has always a more optimistic view than the former. I believe that this is linked to the different approach that academia has on the way results are obtained. Academic research is a creative action and failures or changes, or adaptation are in its essence. You may even decide to wait for the optimal wave, likes a surfer at the seaside. The industrial manager who deals with the production of an asset or the delivery of a service has targets, landmarks, milestones, timing and schedules. This two views are somehow colliding now that the site goes live, but not without some pain on both sides. In fact, a number of issues have quickly deserved more attention than expected.

The industrial partner makes pressure to have precise and timely endpoints, we struggle to set up a calendar despite the heavy internal bureaucracy. In particular, purchases are a nightmare. The logic behind a public tender is based on fairness and transparency, but definitely it is not as quick as it would be in a private business. A lot of details need consideration, nevertheless the risk of complains is always present. The selection of the supplier must be made on the basis of a robust comparison and a solid motivation, and lower price (a generally accepted motivation) may be not the solution you can accept when, for reasons linked to the process, you would need to conclude a direct deal with just one company, which was qualified during the process development. I must admit, it was a very educational experience and a number of solutions we found, have become irreversibly absorbed by the team at the site.

There was another point that proved very challenging along the way: writing SOPs and organizing our activity under a Quality Management System (QMS). Nothing is further from the heart of a researcher, than a bulk of SOPs. The researcher has changes in his DNA, the GMP operator has deviations and change control.

This is to say that the phase in which the GMP was implemented at the site was an uphill effort. The function of the industrial partner was essential, not only to provide the practical support, but to convey the underlying philosophy. However, we decided to build our QMS in line with that of the partner. Importing their QMS and adapting it to our needs is a more correct statement. At that stage, the staff at the site was very small. Two technicians were joined by the core internal responsible persons (QC, QA, and production), who were also involved in production, and myself, as QP and site manager.

Indeed, recruiting people and keeping them on board were the most difficult part of the site organization and management. Hiring the technical staff was not easy: first, there was the issue of radiations exposure (with all the truths and myths behind); second, a permanent staff position could not be offered, but only a temporary contract, linked to the project duration; third, there was the perspective of night shifts, in a future but real scenario.

Unlike a private firm, a public institution cannot modify or adjust the offered contracts, so that special requests can be negotiated and compensated. The position terms are defined by national agreements, which cover salaries, benefits, and available allowances. Thus, when the site had to enter an operational phase, there were a list of problem to solve, including that of recruiting the "right" people, in whom a basic technical skill was paralleled by a good motivation.

The training on SOPs and QMS requires time and effort, this pairs with the training on very peculiar equipment, which are often used under radiation protection protocols.

In other words, it takes months to have an operator who is trained enough and ready to be included in the process. Obviously, when an operator quits, the time to get a valid replacement is terribly long for the site. Nor extra personnel can be hired to face ex ante such a situation: the budget, particularly when production has not yet started, may not be enough to accommodate a larger staff, which could be immediately closing the gap.

A public institution that is dedicated to research probably should never close its doors, and leave researchers the freedom to spend their nights at work! This was my experience in the US, when we could access the chemistry building at any time, even at night: we had the keys! Of course, the security knew who had them, and checked the situation during repeated inspections. Instead, it was incredibly complicated for us, to be present at night! At night, a stringent regulation on safety is enforced at the place where the "Officina farmaceutica" was. After the late inspection of the security, some doors were locked, other had their electronic access code disabled to prevent access to area of possible risk in the absence of the rescue team and other emergency squads. It was necessary to negotiate ad hoc procedures and ensure that more than one operator is present at the same time. The list of names in each shift was then notified to the security beforehand and changes needed a formal approval: they were hard to do at night. The same capillary check applied to trucks, drivers, and any other people reaching the site at night, or outside the normal working hours. It was quite a lot of organizational effort!

The adoption of the night and holiday shift for the team brought forth other complications. The policy and rules on employee working time are defined by a national agreement that, in turn, has practical application rules periodically renewed. The night shift was dismissed from application of the national agreement years ago, simply because nobody was anymore working at night or during holidays. In practice, the possibility for the shifts where there but application "suspended," even if an extra salary was still foreseen to compensate for personnel on shift. It was very complicate to have the article of the national agreement dealing with night and holiday shifts enforced again (the very practical issue was to modify the software handling the presence at work of personnel so that the reading from the attendance recorder, even at night and during holidays, were recognized by the system).

The staff at the site comprised 12 people on average, 6 of them entered the night shifts. My institution has a centralized management of the personnel; it manages some 9000 employee. The "normal" working time ranges from 7 a.m. to 7 p.m., permanence outside this time window should be authorized from time to time. The routine site operation required having people at work from roughly from 9 p.m. to 12 noon of the following day; each day of the week, with the exclusion of Friday and Saturday nights (but Sunday was included). At this point, the negotiation was on multiple fronts: with the central offices, for arranging the shift as to have at least 11 h rest for the operators working at nights, and have their working time accepted by the electronic presence reader; with the personnel, on the maximum number of nights and emergency "call-at-work" in case a replacement was needed; with the security, so that the personnel could be at work even when last minute change modified the list of people on shift or the name of the driver.

In the end, I could find a negotiation with the pool of operators by organizing a rotation of personnel between three "time box": night (10 p.m.–6 a.m.) and day shifts (6 a.m.–2 p.m.), and a standard working time (8 a.m.–4 p.m.). The operator assigned to the standard working time had variable GMP tasks (recording, documentation, supply, etc.) and was not strictly dealing with production; therefore, this time box could be canceled in case of personnel reduction during holidays or illness.

Over the time, another aspect that may become critical is that people get tired and loose motivation. Once again, there was no easy monetary compensation mechanism available. Furthermore, who had an interest in research may become attracted to go back to their previous activity, and having unmotivated personnel on board may be negative to the overall team morale. In particular, this was true for those who were appointed to the positions of responsibility at the production site. Their role at the manufacturing site may not be recognized by academic rules, and this might limit their careers within a public research institution.

It was quite complicate to put all these difficulties on track, but finally the all organization found its way, mostly with the use of SOPs that provide the necessary tutorial and indications on how manage the "atypical" situations we had introduced in our workplace. The approach worked well enough that the institute metabolized the concepts of quality management and got "as a whole" an ISO 9001 certification of its organization. So, in the end, the Officina farmaceutica contributed to open a new chapter inside my institution, a sensitivity on quality which has remained until now!

GMP First Inspection and the Gate to Manufacturing

The inspection by the HA is a thrilling experience, much more than an exam. An "in principle" plan of inspection is communicated by the HA inspector, together with the scheduled date, 1 week in advance to the site visit. The rationale is that you are inspected as you are at the moment, and not after an ad hoc preparation because you knew the inspector was coming. In addition, the progression of the evaluations or the topics inspected can easily deviate from the initial plan and follow actual findings.

It is not wise to delay the visit! Thus, when the notice of inspection arrives, a frenetic activity begins and will last for the entire period of time between the communication of the inspection and the visit. The aim is at double checking the site and of all registrations. The point is: either you are ready or you can delay or even fail the certification.

The visit allows for a complete inspection of the site: in 3/5 days, two to three inspectors analyze the QMS, the VMP, the validation batch records, and any problem that arises during the examination of the records and the documentation. Change control reports, deviations and corrective actions, and emergency reaction (e.g., product recall) receive particular attention. As we later learned, these documents will be even more watchfully perused in the subsequent inspections of the site, when the progression of routine productions yields more "material" to be scrutinized and of a wider nature.

There is a dynamic interaction between the inspectors and the team, and the QP is the direct interlocutor: HA inspection is a full test of the maturation of the GMP knowledge at the site, with theory and practice exams. The inspection can unroll in any possible direction, including having the inspectors participating in the production during a normal night operation: for them … seeing is believing! On the other hand, the thorough attention shown by the inspectors is a crystal clear demonstration to the whole team that the industrial production of drugs must rely on a robust QMS.

The inspection, unless critical deviations are detected, closes with an immediate notification of a variable number of minor or major deviations, that need to be fixed before the certificate of GMP compliance is delivered to the site. Of course, all corrective actions to deviations must be duly documented to the HA before being accepted.

The issue of the certificate is an important milestone for the site and for the contract giver as well, because the site can be formally assigned to the list of manufacturing sites for the specific radiopharmaceutical (a registered FDG in our initial experience). In Europe, the list of third manufacturers that are authorized to produce a proprietary medical product, needs HA clearance. Therefore, when a new manufacturing site is ready to start production, a variation to the officially approved registration dossier of the product must be presented by the MAA holder, and the amended version formally published on the Country Official Journal. It is not an immediate procedure: few months are required before all is fixed. This means there is a time window in which the site and the firm work on the refinement of technical skills and on testing how the communication between the two is functioning.

At this stage, two parallel processes will be tested (event simulation) before the manufacturing starts and real customers are involved: management of shipment, recall, and failures. It is a testing workbench, to prove whether the most operational SOPs at the site are effective or diverge, even in smaller details. The dispatch of FDG or any other short-lived products requires a highly coordinated activity between the site and the carrier. Each package follows a door-to-door transportation, which means a number of containers has to be duly and timely distributed to carriers and controlled for correct destinations.

The logistics can be either managed by the site or assigned to the industrial partner, which is also the Marketing Authorization (MAA) license holder and may have a dedicated service for the management of medicine distribution working in compliance with good distribution practice, an integral part of the GMP. Of course, the site must participate in the process in all cases, keeping contact with the carrier(s), the marketing service of the industrial partner and occasionally the customer(s). It is an orchestra that, at nighttime, plays a music in which a part of the musical score is not completely written, and has to be finalized on the spot.

In fact, only a strong and smoothly organized network of cooperation can support such a complex mechanism, spanning from precursor synthesis to shipment of finished product door-to-door in a few hour time window.

A Personal Story: The Night at Work

The first times the site operates during the night, it is a quite new experience. Nobody is around, apart the team on duty and the security officer, who looks at you as one more problem among others.

The GMP certificate was achieved, the additional paperwork to start the formal production with our industrial partner was ongoing, all proceeded smoothly. While waiting for starting the industrial production, we were producing on the basis of the FDG monograph of the European Pharmacopoeia and shipping to the general hospital in Pisa. Some 20–30 min drive away, door-to-door. The "real production" would start at 4 a.m., with the bombardment of the ^{18}O-enriched water at the cyclotron to prepare the precursor radionuclide, ^{18}F-fluoride. However, the night shift, as agreed with the personnel office in Rome, had to begin earlier; therefore, we added an earlier extra run, just an exercise and a test. The product was prepared and tested according to all specs, and remained available, in case a back-up was necessary to support anyone had a problem in the "FDG manufacturing Club." Indeed, we shipped the product we had available more than one time; and we felt very proud to help patients and colleagues (regardless whether they belonged to cognate or com petitor's sites).

Surprisingly we had a little more difficulty with shipments to the closest locations than those farther away; this is due to the fact that analytical checks on the batch are normally carried out while the product is on its way to the customer(s). In fact, in some cases, the analytical controls and the checks needed for releasing the product for use lasted almost the time necessary for the product to reach the customer: normally we had more time and the certificate was received by the customer before the product delivery. Only the transmission of the certificate of analysis would have cleared the use of the product, which delivered "in bond."

When we shipped to Pisa, i.e., downtown, the product reached the user before we could complete all tests and have the possibility of sending the releasing certification. Therefore we had to ask them not to open the shipment bag (a sealed plastic Type A ADR container) before we had sent the certificate of analysis by fax.

The personnel at the user's site was duly instructed, nevertheless we always remained concerned of a possible misuse until the product was released; therefore, as an additional precaution, we asked the driver to participate in this safety procedure by keeping the product under his custody until the release was effectively communicated to the hospital, as demonstrated by showing the driver the releasing documentation. The inspectors of the HA (and perhaps ourselves too) were skeptical that the doctors would have the patience to wait our green light, when the product was in their hands. Therefore, we used this "escamotage," just to be on the safe side. In fact, should the customer had opened the bag and the batch had not passed the quality control, a recall would be necessary. This is an emergency procedure that is activated in the event that the product presents a defect and is on the market: i.e., has reached the nuclear medicine unit and is used (or prepared for use) by the customer. The recall is a complex procedure that involves the HA, in which we (and any other parties or partners involved) had to explain how a defective product could reach the market: a scary movie not to watch and above all to play as main characters!

6.4   Now That You Got the Horse, Ride It

Navigating the Routine Production

With an expiry time of about 10–12 h, the FDG is produced, delivered, and promptly used; obviously, no stock is available. What takes weeks in a conventional pharmaceutical industry is frantically compressed at a FDG manufacturing site, including all controls and approvals that pend over a drug. In a typical day, the QP has to review more than one batch, and release each batch for use before it could be administered to patients at the customer hospitals, only a few hours since the end of production (EOP): timeliness has never been more precious. It is a very high pressure on all the team and the organization.

The time pressure is so high that many controls are performed on a sample of the finished product while the packaged vials are traveling to their final destination. This is again a peculiarity of this kind of products and is the major stress on the team at the site. Formally, SOPs have been established to ensure the product is not available to the users before the quality control is passed and product approved for use. The team at the site has no spare time to rest, in particular when multiple runs are scheduled, which means a full cycle of operation (a 2–3 h duration, from cyclotron operation, synthesis, QC, and shipment) is repeated up to four times a day.

A full day of operations requires that the night and day shifts exchange smoothly even when a run is ongoing. Of course, a full mastering of the process is necessary by any member of the team and in each of the "departments," namely production and quality control. In fact, the personnel assigned to one department cannot operate in the other, to maintain full independency between who produces and who checks and controls. At least, this is the rule in many EU countries.

If the normal routine is complicate, the emergency situations are obviously worst. The most frequent emergency situation at a PET radiopharmaceutical manufacturing site deals with product availability: when the process fails, or the amount produced is insufficient with respect to the production plan, the customers risk not to receive the tracer, and have their patient untreated.

In these cases, a frantic activity is started at the site, in cooperation with the personnel at the industrial partner's office, to identify the best strategy to solve or mitigate the emergency and, as soon as possible, put a solution on track. When product yield is low, a first attempt is to rebalance the distribution and optimize the destinations (i.e., have as much activity as possible reaching the hospitals); meanwhile, the equipment is restored to be ready for an additional run.

When the production completely fails, unless a critical hardware fault is the root cause, new runs are added to the production plan in order to fulfill the customer's requirements, although with a shipment later than the planned schedule. Almost in parallel, the possibility of receiving a back-up is also explored, so that any redundant product existing at other sites in the network, or even at competitor sites,[5] can be supplied to the customers that would otherwise miss their shipment.

However, even the normal routine is not an easy story: in the real life, when the site is running, the team and the QP remain alerted until the product has left the site, and any ring of the phone in the night leave many people with a sinking heart.

A Personal Story: The Night Stress

After some years of operations, the site runs full engines: three to four batches are produced each day.

The team has sometime experienced low yield and even failing productions, and the unpleasant situations that follow: stress inside the team, complaints from the industrial partner, and sometime from those customers who, having got your coordinates, decide to express directly their disappointment to you, without the interposition of the partner customer service.

When delivery are not finalized as expected, it is a very frustrating experience, of course complains do not please, but the worst is that the failure produced discomfort and troubles to patients. Therefore, as early as possible, as soon as a risk of failure is deemed possible, an emergency call leave the site to reach the responsible for emergency management of the partner. A breathless effort starts to fix the situation: the production run may be completely repeated, or integrated by a new one in case of low yield. It can even be replaced by a back-up delivery coming from another source, including a competitor. The goal is not to leave the patients untreated.

It is a turmoil of calls, checks, reorganization of planned transports, change of documents. There is a choral effort by all on the stage: the site QP and managers, the team on duty, the partner, the drivers that were alerted to carry the product. The customer is contacted as soon as he is reachable and alerted that may receive a delayed delivery of the product or, in the worst case, no product at all. Unfortunately, even if the production problem occurs in the heart of the night, the personnel at the hospital learn this situation only when they come on duty and open the diagnostic service, that is when patients are already there or are about to arrive.

Rarely, the reactions to these problems are calm and sympathetic to your efforts. To make things even worst, all this occurs at night, and even your family may not be willing to support you beyond a certain limit (or nights with so many calls).

---

[5] When mitigation actions are not sufficient and the failure is critically hampering that the tracer reaches all patients, destinations can be covered by competitors. Patients cannot remain untreated as much as possible.

Notwithstanding the efforts and the attention, it is extremely challenging the management in 2 h of a whole pharmaceutical process, ranging from the production of the radioactive nuclide at the cyclotron to the shipment of the radiotracer. The variables to keep under control are really a lot and failures are hard to fix in such a short time frame, while the product decrease at a rate of roughly 8% every 10 min, with very high level of radioactivity so that you cannot open the hot cells or the containers, and, the product still has to reach its final destination by truck!

Moreover, the effects of incomplete of failed productions are not limited to a night storm, but extend to the following days. In fact, these events are included in the pool of process deviations, and the investigation is started immediately to identify the root cause and the consequent corrective/preventive actions. The more you get data on the process, and on failures as well, the higher is the predictive value of trending analysis to identify the weaker steps of the process in your hands. Key Performance Indicator (KPI) was identified and used to do this at best, and adapted to any new critical situation encountered. The evaluation of Out of Trend (OOT) applied to KPIs and analytical data helps in the definition of preventive actions that can mitigate the incidence of failures and Out of Specifications (OOS), i.e., the negative outcome of the process. The intensification of global maintenance, but in particular its focusing on specific harnesses (the root cause from the deviation analysis) was successful in limiting technical and instrumental failures. In parallel, a set of spare parts, those emerging as critical because of frequent ruptures or poor availability, can be stored on-site so that time is saved upon correction of broken/defective equipment.

Indeed, the intrinsic limits of a microgram synthesis still persist, and 100% success unreachable. As a QP, I have released many thousands of batches, over the many years of operation of the site, and I have thought many times about the Indian headdress I saw during my first FDG synthesis, long time ago at Brookhaven.

There have been quite harsh and stressing moments, when the root cause was hard to find. It is the moment in which some people, in particular the QP and the site manager (sometimes the same person, as in my experience), who first took the idea of starting a manufacturing site, may be asking themselves why they did not think it could be a small ship in troubled waters. The memory of calmer moments (the in-house radiopharmacy) comes to the minds, also to the technical staff. From time to time, the managers and even the director have to keep the team on track and provide them with motivation and support. However, the awareness that this project provided a valuable service to patients and that new research projects have obtained financial support that would otherwise be impossible at my institute have been strong motivations and a source of pride for me and my team.

An Overall, Personal Consideration

After years of operation as third-party manufacturer at a site organized inside a public institution, a question may come up: was it worthwhile? Would I do it again?

There are different ways to answer the question. Obviously, some personal, base-line considerations should be recalled, in the chronological order in which they consolidated. Firstly, I was pushed by the drive to find new areas of diffusion and use of science (chemistry) and technology applied to medicine; second, I developed the passionate attention to helping patients and promoting a social benefit on health-care through my work. Other aspects, such as the development of innovative part-nerships and the economic revenue, were also important, but moved in parallel to the previous ones.

Let me start from the comparison of my initial experience, as the responsible of an in-house PET radiopharmacy, and what came later, as QP and site manager of an industrial radiopharmaceutical manufacturing site, at least as they are in my percep-tion. Both have full dignity and also positive and negative aspects that are strictly related to the intrinsic organization and technical complexity of each solution (Fig. 6.3 Pro and cons in-house radiopharmacy and industrial radiopharmaceuti-cal site).

The decision to move on to one direction or the other can find a rationale justifi-cation in both cases, but changing the direction from an in-house radiopharmacy to an industrial set up depends on some basic points. At many public hospitals and clinical research centers, a cutting point discussion is whether having an in-house PET radiopharmacy (with cyclotron and radiochemistry) is (or continues to be) a viable decision or not. Some points play a major role: the existing level of

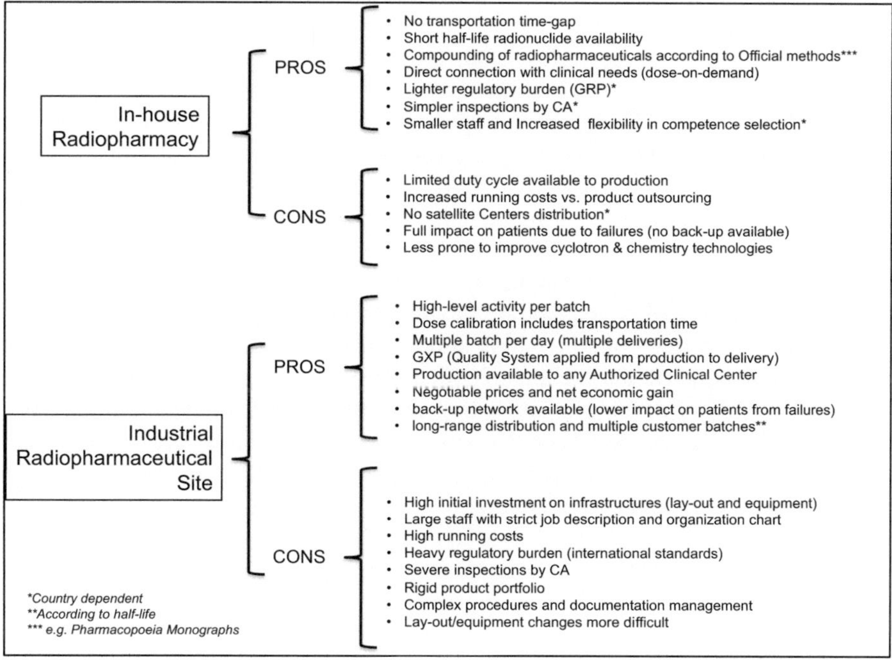

Fig. 6.3 Pro and cons in-house radiopharmacy and industrial radiopharmaceutical site

technology and staff, the price and availability of commercial PET tracers, the running costs of the site, and the sustainability of the on-site use of tracers other than those commercially available.

How these aspects contribute to the decision is based on the situation where a fairly large and well-organized setup remains compressed at its peak performance and needs to make a decision whether to collapse on itself (commercial tracer may become cheaper than internal production costs) or moving on to the next level, which includes a broader exposure in terms of tracers and prospective applications. In addition to a mere economic decision, the last point also includes the presence of clinical and/or research projects that may provide economic or scientific justification for keeping the installation alive, but may need a more stable influx of funding, larger than those available through purely competitive calls. Of course, these are quite coarse criteria of analysis, and local situations or conditions may give raise to different decisions and conclusions.

A private owned in-house PET radiopharmacy stays outside this dynamics. In general, it is opened, and it keeps existing, if it passes a sustainability analysis. The running costs (cyclotron, radiochemistry, and staff) need to be balanced by the revenues from the internal diagnostic department; in fact, many PET/CT scanners are usually pooled at the same location in private diagnostic center, so that an intense service can be provided to both private outpatients and patients referred to the center, via contracts and agreements, by the public healthcare system.

However, once the decision is taken, the transition from a public in-house PET radiopharmacy to a GMP industrial site is a challenging task, because a connection has to be found between public and private worlds, that normally operates with different philosophy and targets. First of all, a necessary step is to understand what an industrial processes is in particular the meaning of manufacturing and distributing a product to customers. The first point to ponder is that it will be adopted one specific process, that has been developed in details and has followed a complex and meticulous track of approval. The "creative" drive should be refrained on that. The second is that the industrial manufacturing includes the commitment to deliver the product to the specified customers, at the stated time, having the required specifications, starting from large, multiple batches of products labeled with short-lived radionuclides. This is a critical issue, and one should be well aware of it beforehand.

That said, a starting point is to look at how the project can position itself within the accessible market and try to identify what can help or jeopardize its success. The industrial world has developed interesting tools to separate the potential variables that may affect a project and have the best chance to ponder possible resources and risks as well. An example of a SWOT[6] analysis, conducted on the basis of my personal experience, is summarized in Fig. 6.4.

Of course, the single items reported in the table may vary from country to country, may differ on the basis of the organization of the public entity taking the effort, and be strongly influenced by local or national situations as well.

---

[6] Strengths, weaknesses, opportunities, and threats

Strengths	**Weaknesses**
• Infrastructures (existing laboratories & high-cost equipment with medium/low duty-cycle, availability of general purpose instrumental/technical resources) • Qualified personnel with background competence on selected products (science & technology) • Organization resilience (large scale Public Body) • Scientific & technological advice/support from nearby high-tech assets (research infrastructures) • Low general costs (shared) in multiple research group Institutions): power supply, air conditioning, ICT, .. • Low costs of general services (usually shared or centralized in large multi-spoke National Institutions): Administration, Maintenance, Safety, Security, Personnel office, ... • Fiscal regime of public Bodies • Institutional reputation • Flexible mark-up on manufacturing costs • Availability of competitive PET manufacturing slots (night time and/or during the day)	• Internal bureaucracy & time-to-decision (e.g. for structure like in some National Organization operating with a Central hub & peripheral spokes) • Public Administration rules (multiple authorizations on building modifications, tender rules on acquisition of outsourced services, buying of supply and equipment) • Risks of delayed time-to-target • Personnel management (shift organization and rigid competence grids) • Employee reluctance to timetable flexibility (extra-time, night-time) • Quality Management System absent/far from necessary • Lack of marketing, commercial and customer care experience • Need for logistic support • Difficult to agree on an "exclusive" collaboration, based on specific class of radiotracers • Possible long reaction-time in case of decrease of manufacturing reliability
Opportunities	**Threats**
• Starting of a new Contract Research Organization Service working on the general pharmaceutical market. • Revenues free from obligations and boundary conditions (e.g. as it is with funds/grants derived from supranational Entities) and perspective of accessing upgraded/innovative equipment (reinforce research on basic sciences, high-risk innovation & technologies) • Acquisition of QMS skills and application of high quality standards (GMP/GLP) to promote in-house research products and third-party/industry cooperation in pre-marketing activity (join programs): ad hoc financial support channels • Transition of activity towards an independent Company access to funding for the promotion of high-tech start-up/enterprises): • to gain impact on Regional/National Healthcare market (privileged player role) • Synergy with internal clinical departments to perform Phase 1, 2 or 3 trials	• Volatility of market (product trend & turn over based on medical science achievements) • Market dimensions (additional player effect) • Product portfolio (numerosity, product priority, new developments) • Changes on national Healthcare strategy

Fig. 6.4 SWOT analysis

The more accurate and complete is the filling of the table, the more it will efficiently guide the decision, which depends upon the ranking of the risks (internal weaknesses and external threats), and the evaluation whether and how much they are balanced by the positive resources and perspectives (internal strengths, external opportunities).

Some of the items in the lists of Fig. 6.4 may deserve further comment, because their direct impact on the project may not be clear enough. The "organization resilience" can be a positive resource because a public entity may benefit from compensative actions (relocation of human and economic resources) much more easily than a private firm, which usually move on a predetermined budget and anticipated appropriations.

It was already mentioned that full-staff positions may be considered outside the budget of the peripheric manufacturing site and remain charged on general, central funds allocated for permanent personnel and general costs. The greater the size of the public body, the larger the budget allocation will be to items dedicated to general services, which can be distributed to the peripheral units as gross amounts regardless of the specific destination. This means that costs for general services will be shared by all institutes (Departments or Divisions) and each group, including the manufacturing site, will only pay for a fraction of it. A similar benefit can derive also from the fact that a public institution does not have commerce as its primary commitments. This translates in a different fiscal regime (simplified) and less stringent constraints in price mark-ups.

This "Strengths" should not be spoiled by internal limitations, such as bureaucracy (on both paperwork and personnel) and organizational gaps. A private organization can much more efficiently decide on strategies and quickly achieve practical goals, or simply assign resources to tasks. However, just for this reason, the public body can receive a positive support from the industrial partner that may convey resources to cover these gaps within a partnership agreement.

The external context is indeed variable. It may change suddenly and needs a careful, ongoing attention along the entire evolution of the project. However, variations may come unexpectedly. For instance, the amount paid per exam by the public healthcare system can be reduced to respond to national healthcare budget restrictions. This can change the revenue to the contract giver and, in cascade, affect the payment per lot of radiotracer, as well as the number of batches required. The contract giver may also withdraw, because of a change in its market strategy, and leave the third party without a regulatory coverage (the contract giver is also the owner of the marketing license of the product(s) manufacturer at the site) or with a limited coverage in case the site has more partners in its portfolio.

All these issue should be considered in case of transition toward an industrial-like organization, and used in the planning of the strategy and in the writing of the contract(s). The wiser is the analysis, the higher will be the chance of having a profitable and successfully experience.

Over the years, the opportunity of having a more flexible asset was the winning solution. In fact, the partnership based on a single-partner, single-product agreement exposes the public site to a higher risk of negative influence from both the market (when the target product should exit the market or receive a reduced interest by the medical community) or any possible adverse decision coming from the industrial partner (e.g., a change in the approach to the market).

Of course, keeping a site alive with diversified productions increases costs and yields higher documentation burden. Therefore, a careful evolution must be given not only to the technical staff concerned with the production processes, but also to the organization of the supporting activity, such as clerical work, maintenance, and administration. In practice, a "business unit" should be established, in which a full confidence is reached on budget management.

If this target is achieved, the site—provided the infrastructure allows it—can even resume its original commitments with research and explore the possibility to add, on the side of the third-party manufacturing, the services provided by a contract research organization in the development of new tracers and radiopharmaceuticals.

6.5 Conclusions

Now, the site has been existing since about two decades. Different contract givers have changed, and different products have been produced over the time. Indeed, there was an intense learning curve on many issues, and very challenging and complicate periods have passed. It was a very intense experience, combining human and professional growth. Many thousands of FDG batches have been produced, and even more patients treated.

This is happening even today that I retired, and this makes me very happy.

Final Greeting

We hope that the reader has had the opportunity to explore aspects of PET production that can help him in daily practice.

From the very beginning, this has been our goal, preferring to mention the "not to do" things over the "to do" things.

The literature, in our opinion, is extremely rich in technical, regulatory, and clinical references about PET radiopharmaceuticals, but it seemed a bit lacking of basic field experiences.

In addition to investigating aspects related to the operational management of PET sites and the relationship with the subcontractor that I had already dealt with in the first volume, through the experiences of Maria Carmela and Piero, we had the opportunity to open a window on two new worlds.

The first, that of quality assurance, was described by a professional who for 30 years has struggled to try to align the peculiarity of the radiopharmaceutical, regulatory requirements, and sustainability in the management of the system. The latter point is not taken for granted, in light of the reduced availability of personnel and the unpredictability of the PET processes that can tempt the company to prioritize the most urgent operational aspects rather than those of an accurate analysis and traceability of the same.

I am convinced that anyone, whether it is in a new or established PET reality, can benefit enormously from what Maria Carmela described in her chapter.

The second, that of the development of a PET facility in a hospital is a situation that I have often encountered, at least in terms of feasibility requests, in the context of my professional life. However, I have often realized that my indications, easily applicable in the context of a private company, clashed with the specific constraints operating in a public structure. Piero has managed to combine research activities on the development of PET tracers and related clinical practices with production activities, and has finalized outsourcing collaborations during 40 years of activity in the public sector. Therefore, I think that his experience can help public professionals who intend to start an industrial PET experience to analyze the project, in terms of possible problems and operational solutions.

Thanks to everyone who had the patience and perseverance to read this text.

Glossary

Back-up Supply of the radiopharmaceutical agreed in case of temporary interruptions of production due to ordinary or extraordinary maintenance.

CMO (Contract Manufacturing Organization), subcontractor.

ISO 9001 A voluntary certification that defines the minimal requirements for the quality management system (QMS) and monitor its efficacy.

Label Drug labels identify drug contents and state specific instructions or warnings for administration, storage, and disposal.

Pharmaceutical manufacturing site Plant which, following specific authorizations, is able to carry out all operations from purchase of raw materials and products, to production, quality control, release, storage, and distribution of approved medicines.

PET pharmaceutical manufacturing site Pharmaceutical manufacturing site specifically dedicated to the production of radiopharmaceuticals for PET diagnostics.

PET Acronym for "Positron Emission Tomography." In vivo nuclear medicine diagnostic method based on the detection of the distribution of positron emitting radionuclides.

Procedural system List of documents, reviewed and approved by the responsible company functions, which regulate all the activities of the pharmaceutical manufacturing site.

QA Quality Assurance. Team/Department that in a pharmaceutical company guarantees that the principles and standards of Good Manufacturing Practices (GMPs) are always applied consistently.

Radiation protection Discipline that studies methods to safeguard humans from the biological damage that radiation of any kind can cause.

Research and Development (R&D) In here, it refers to the pharmaceutical research and development of new medicines and covers a variety of activities, including discovering and testing new drugs, developing incremental innovations such as product extensions, and clinical testing for safety-monitoring or marketing purposes.

SPECT Acronym for "Single Photon Emission Computed Tomography." In vivo nuclear medicine diagnostic method based on the detection of the distribution of gamma emitting radionuclides.

Tele manipulator A remote controlled device for performing the manipulation of objects that are too heavy, dangerous, small, or difficult to handle directly.

Further Reading

Calabria F, Schillaci O. Radiopharmaceuticals. Springer; 2020.
EudraLex. Good Manufacturing Practice (GMP) guidelines, vol 4.
Glaudemans AWJM, Medema J, van Zanten AK, Dierckx RAJO. Quality in nuclear medicine. Springer; 2017.
La LG. Qualità nella preparazione dei radiofarmaci. Springer; 2011.
Lewis JS, Windhorst AD, Zeglis BM. Radiopharmaceutical chemistry. Springer, 2019.
Pecorale A. Essence of the PET radiopharmaceutical business: a practical guide. Springer; 2022.
Saha GB. Basics of PET imaging, physics, chemistry, and regulations. Springer; 2016.
Stöcklin G, Pike VW. Radiopharmaceuticals for positron emission tomography. Springer; 1993.
Volterrani D, Erba PA, Carrio I, Strauss HW, Mariani G. Nuclear medicine textbook. Springer; 2019.

A. Pecorale et al., *PET Radiopharmaceutical Business*,
https://doi.org/10.1007/978-3-031-51908-6